SCI PUBLICATION 264

Wind-moment Design of Unbraced Composite Frames

J S HENSMAN BEng, MPhil
A G J WAY MEng

Published by:

The Steel Construction Institute
Silwood Park
Ascot
Berkshire SL5 7QN

Tel: 01344 623345
Fax: 01344 622944

© 2000 The Steel Construction Institute

Apart from any fair dealing for the purposes of research or private study or criticism or review, as permitted under the Copyright Designs and Patents Act, 1988, this publication may not be reproduced, stored or transmitted, in any form or by any means, without the prior permission in writing of the publishers, or in the case of reprographic reproduction only in accordance with the terms of the licences issued by the UK Copyright Licensing Agency, or in accordance with the terms of licences issued by the appropriate Reproduction Rights Organisation outside the UK.

Enquiries concerning reproduction outside the terms stated here should be sent to the publishers, The Steel Construction Institute, at the address given on the title page.

Although care has been taken to ensure, to the best of our knowledge, that all data and information contained herein are accurate to the extent that they relate to either matters of fact or accepted practice or matters of opinion at the time of publication, The Steel Construction Institute, the authors and the reviewers assume no responsibility for any errors in or misinterpretations of such data and/or information or any loss or damage arising from or related to their use.

Publications supplied to the Members of the Institute at a discount are not for resale by them.

Publication Number: SCI-P264

ISBN 1 85942 114 8

British Library Cataloguing-in-Publication Data.
A catalogue record for this book is available from the British Library.

FOREWORD

This publication has been prepared by Mr Jason Hensman of Caunton Engineering and Mr Andrew Way of The Steel Construction Institute. The background analytical work leading to the publication was carried out by Mr J Hensman whilst studying at the University of Nottingham. The assistance of Dr Graham H Couchman (The Steel Construction Institute) in developing the publication is acknowledged.

SCI publications related to the wind-moment method are:

Wind-moment design for unbraced frames (SCI P082, 1991, superseded by SCI P263)

Wind-moment design of low rise frames (SCI P263, 2000, replacement for SCI P082)

Joints in steel construction: Moment connections (SCI P207, 1995, SCI/BCSA)

Funding for the project was gratefully received from the Department of the Environment, Transport and the Regions (DETR) and Corus (formerly British Steel).

CONTENTS

                                                            Page No.

SUMMARY			vii
1	INTRODUCTION		1
	1.1	Benefits of the wind-moment method	1
	1.2	Benefits of composite construction	1
	1.3	Scope	2
2	PRINCIPLES OF WIND-MOMENT DESIGN		6
	2.1	Design method	6
3	CONNECTIONS		10
	3.1	Connection classification	10
	3.2	Beam-to-column connections	10
	3.3	Standard connections	12
	3.4	Column bases	13
4	DESIGN FOR THE ULTIMATE LIMIT STATE		14
	4.1	Global analysis	14
	4.2	Design of beams	15
	4.3	Design of columns	16
	4.4	Design of connections	17
5	DESIGN FOR SERVICEABILITY LIMIT STATE		19
	5.1	General	19
	5.2	Sway prediction	19
6	REFERENCES		23
APPENDIX A:	Portal method of analysis		27
APPENDIX B:	Connection detailing		31
APPENDIX C:	Capacity Tables		37
APPENDIX D:	Worked Example		51
			51

SUMMARY

This publication presents procedures for the design of wind-moment composite frames in accordance with BS 5950-1 and BS 59550-3. In this method of design, the frame is made statically determinate by treating the connections as *pinned* under vertical loads and *fixed* under horizontal loads (with certain assumed points of zero moment). The publication gives design procedures for frames (with composite beams, slabs and connections) that are braced in the minor axis direction. The limitations of the method are explained. In particular, it should be noted that the method is only recommended for low-rise frames up to four storeys high.

In addition to design procedures for the ultimate and serviceability limit states, fully worked design example is presented. The publication also reproduces the resistance tables for standard wind-moment composite connections taken from SCI/BCSA publication *Joints in steel construction: Composite connections*. These connections use steel reinforcement, flush end plates and grade 8.8 M20 or M24 bolts, and achieve sufficient rotation capacity by ensuring that the moment resistance is not governed by local concrete crushing or bolt or weld failure.

Dimensionnement sous moments dus au vent de cadres composites non contreventes

Resumé

Cette publication présente des procédures de dimensionnement de cadres composites sous moments dus au vent qui suivent les normes BS 5950-1 et BS 59550-3. Dans cette méthode de dimensionnement, le cadre est rendu statiquement déterminé en considérant les assemblages comme articulés sous les charges verticales et encastrés sous les sollicitations horizontales (avec certains points où on suppose les moments nuls). La publication donne des procédures de dimensionnement pour les cadres (avec poutres, dalles et assemblages composites) contreventés seulement dans la direction de l'axe faible d'inertie. Les limites de la méthode sont expliquées. En particulier, on doit noter que la méthode est seulement recommandée pour les cadres ne présentant pas plus de quatre étages.

En plus des procédures de dimensionnement pour les états limites ultimes et de service, des exemples complets de dimensionnement sont présentés. La publication reproduit également les tables de résistance pour des assemblages composites standardisés repris de la publication SCI/BCSA "Joints in steel construction : composite connections".

Ces assemblages utilisent des plaques d'about ou non débordantes avec boulons de type 8.8 M20 ou M24 et assurent une capacité de rotation suffisante en imposant que la résistance en flexion ne soit pas conditionnée par l'écrasement local du béton ou par une rupture d'un boulon ou d'une soudure.

"Wind-Moment-Berechnung" verschieblicher Rahmen aus Verbundbauteilen

Zusammenfassung

Diese Publikation präsentiert Vorgehensweisen für die Berechnung von Rahmen aus Verbundbauteilen unter Einwirkung von Momenten infolge Windlasten ("wind-moment composite frames") nach BS 5950-1 und BS 5950-3. Bei dieser Methode wird das Tragwerk statisch bestimmt gemacht durch Annahme von gelenkigen Verbindungen unter vertikalen Lasten und biegesteifer Verbindungen unter horizontalen Lasten (mit gewissen angenommenen Momenten-Nullpunkten). Die Publikation zeigt Berechnungsweisen auf für Rahmen (mit Verbundträgern, -decken und Verbindungen) die bezüglich der schwachen Achse unverschieblich sind. Die Grenzen der Methode werden erläutert. Besonders sollte beachtet werden, daß die Methode nur für Rahmen geringer Höhe mit bis zu vier Geschossen empfohlen wird.

Zusätzlich zu den Berechnungsmethoden im Grenzzustand der Tragfähigkeit und Gebrauchstauglichkeit werden Berechnungsbeispiele vorgestellt. Die Veröffentlichung reproduziert auch die Tabellen für Standard-Verbindungen im Verbundbau unter Einwirkung von Momenten infolge Windlasten aus der SCI/BCSA-Publikation Verbindugen im Stahlbau:Verbindungen im Verbundbau. Diese Verbindungen berücksichtigen die Bewehrung, bündige Stirnplatten mit Schrauben M20 oder M24 der Güte 8.8 und weisen ausreichende Rotationskapazität auf, ohne lokales Betonversagen, Schrauben- oder Schweißnahtversagen.

Dimensionamiento de porticos sin arriostramientos frente a momentos de viento

Resumen

Esta publicación presenta métodos para el dimensionamiento a carga de viento de pórticos mixtos de acuerdo con BS5950-1 y BS59550-3. En este método de cálculo el pórtico se convierte en isostático considerando las uniones como articuladas ante cargas verticales y fijas para cargas horizontales (con ciertos puntos donde se supone momento nulo). La obra da métodos para pórticos (con vigas mixtas, placas y uniones) arriostradas en dirección del eje menor, explicándose sus limitaciones. En particular debe hacerse notar que la aplicación del método se limita a edificios de menos de 4 alturas.

Además de métodos de proyecto para los estados límites de servicio y último, se presentan ejemplos desarrollados. También se reproducen tablas con resistencias de uniones mixtas tipo, tomadas de la publicación. "Uniones en estructuras de acero. Uniones mixtas" de SCI/BCSA.

Estas uniones usan armaduras de acero, platabandas y tornillos de calidad 8.8 M20 o M24 y consiguen una capacidad de rotación suficiente al garantizar que el momento resistente no está controlado por aplastamiento local del hormigón o por roturas de soldaduras o tornillos

Progettazione alle azioni orizzontali di telai non controventati composti

Sommario

Questa pubblicazione presenta gli approcci per la progettazione alle azioni orizzontali di telai composti secondo le normative BS5950-1 e BS5950-3. In accordo a questo metodo progettuale, il sistema intelaiato viene considerato isostatico, schematizzando i collegamenti come cerniere in presenza di carichi verticali e come nodi rigidi in presenza delle azioni orizzontali (con prefissate zone in cui l'azione flettente viene ipotizzata nulla). La pubblicazione propone le procedure progettuali specifiche per i sistemi intelaiati (con travi composte, solette e collegamenti) che sono controventati nella direzione di minore rigidezza della colonna.

Sono descritte le limitazioni del metodo e, in particolare, viene precisato che questo è applicabile soltanto ad edifici di altezza modesta, fino a un numero massimo di quattro piani.

A corredo della trattazione progettuale proposta e riferita agli stati limite, sia di servizio sia ultimi, vengono presentati anche quattro esempi progettuali completi. La pubblicazione riporta anche le tabelle di resistenza per le più tradizionali tipologie di collegamenti per sistemi composti, già proposte nella pubblicazione SCI/BCSA 'I giunti in strutture metallliche:i collegamenti composti'. Questi collegamenti, che vengono realizzati mediante barre di armatura in acciaio, giunti flangiati con piatti in spessore di trave o estesi oltre l'altezza di trave e bulloni di diametro 20mm o 24mm e classe di resistenza 8,8, consentono di ottenere un'adeguata capacità rotazionale garantendo che la resistenza flessionale del nodo non sia condizionata dalla fessurazione locale del calcestruzzo o dalla crisi nei bulloni o nelle saldature.

Vindmoment dimensionering av oförstärkta sammansatta ramar

Sammanfattning

Denna publikation presenterar ett metod för dimensionering av vindmoment i sammansatta ramar i enlighet med brittiska standarderna 5950-1 och 59550-3. I denna metod är ramen är gjord statiskt bestämbar genom att behandla alla knutpunkter som fritt upplagda för vertikala laster och alla knutpunkter som fast inspända för horisontella laster. (där vissa punkter antas ha noll moment). I publikationen finns dimensioneringsmetoder för ramar (med sammansatta balkar, plattor och knutpunkter) som är förankrade i den mindre axelns riktning. Begränsningarna med denna metod är förklarade. Det är värt att notera att denna metod enbart rekommenderas för ramverk på upp till 4 våningar.

Som tillägg till dimensioneringsmetoden presenteras helt genomarbetade exempel för att underlätta användbarheten. Publikationen redovisar även tabeller för standard vindmoment för sammansatta knutpunkter tagna från SCI/BCSAs publikation "Joints in steel construction: Composite connections". Dessa knutpunkter innehåller stålförstärkningar med glatta eller förlängda ändplåtar och klass 8.8 M20 eller M24 bultar. Knutpunkterna erhåller tillräcklig rotationskapacitet genom att det är säkerställt att böjmotståndet inte begränsas av lokala tryckbrott, bultbrott eller svetsbrott.

1 INTRODUCTION

When a steel frame is unbraced, an established technique to provide resistance against wind loading is to rely on the rotational stiffness of the beam-to-column connections; under vertical load the connections are assumed to be nominally pinned. This design philosophy has become known as the *wind-moment method* or *wind-connection method* and is used extensively in the UK and North America.

The wind-moment method is described in the SCI publication *Wind-moment design of low rise frames*[1] for bare steel frames, and this publication is an extension of that guidance for composite frames.

The frames covered by this publication comprise composite steel and concrete beams, composite beam-to-column connections, and steel columns. The procedures given in this publication are consistent with the design rules given in BS 5950-1[2] and BS 5950-3, Section 3.1[3].

1.1 Benefits of the wind-moment method

One of the main advantages of the wind-moment method from the designer's point of view is its simplicity, as is explained in Reference 1[1]. The internal moments and forces are not dependent on the relative stiffnesses of the frame members because the frame is treated as statically determinate (see Section 2). The need for an iterative analysis and design procedure, which usually makes the use of software virtually essential, is therefore avoided.

From a construction viewpoint, the major advantage of wind-moment frames is the relative simplicity of the steelwork when compared with fully rigid construction. Much of the work carried out by steelwork contractors is concerned with making the connections, and it has been estimated that the fabrication and workshop handling costs associated with the connections can be as high as 50% of the total cost of the erected steelwork.

1.2 Benefits of composite construction

The general benefits of composite construction, as given in the SCI publication *Design of composite slabs and beams with steel decking*[4], are:

- Reduction in weight of the steel beams by 30 to 50% when compared with non-composite construction, leading to a significant reduction in frame cost.

- Increased stiffness of the floor construction giving better serviceability performance.

- Longer spans for a given beam depth or, alternatively, shallower construction for the same span.

- Good load resistance and robustness.

- Diaphragm action through the floor slab, eliminating plan bracing and providing direct load transfer to vertical bracing or core walls.

- Rapid construction of both the steel framework and concrete slabs is possible by using the decking as both a working platform and permanent formwork.

The procedures given in this guide enable designers to combine the benefits of the wind-moment method and composite construction.

1.3 Scope

The recommendations given in this design guide have been formulated using knowledge gained from an extensive background study[5]. The limits of application have, where appropriate, been based on the existing wind-moment method limits defined in Reference 1 and *Wind-moment design of unbraced frames*[6]. It should also be recognised that the scope is dependent on the ranges considered in the background study, and hence the method has only been validated within the given limits.

1.3.1 Frame proportions

For composite wind-moment design, the geometry of the frames should be within the limits defined in Table 1.1. Frames may have more than four bays, but any additional bays (over and above four) should not be considered to participate in resisting the applied wind load. When adding non-active bays, the designer should ensure that the notional horizontal loads do not become critical.

Table 1.1 *Frame geometry limits*

	Minimum	Maximum
Number of storeys	2	4
Number of bays	2	4
Bay width (m)	6.0	12.0
Bottom storey height (m)	4.5	6.0
Storey height elsewhere (m)	3.5	5.0
Bay width: storey height (bottom storey)	1.33	2.67
Bay width: storey height (above bottom storey)	1.33	3.43
Greatest bay width: Smallest bay width	1	1.5

Frames designed using this design guide must also display the following features:

- The frame must consist primarily of horizontal beams and vertical columns (Figure 1.1).
- The width of each bay must be constant over the height of the frame, except that the columns may terminate at the top floor to allow an open-plan top storey (Figure 1.2).
- Frames must be effectively braced against out-of-plane sway at each floor level and at the roof level (i.e. major axis sway frames only).
- Beam grids must be consistent with one of the layouts shown in Figure 1.3.
- Composite slabs should span in directions as shown in Figure 1.3.
- Sections should be orientated in such a way that loads in the plane of the frame tend to cause bending about the major axis (i.e. major axis sway frames only).

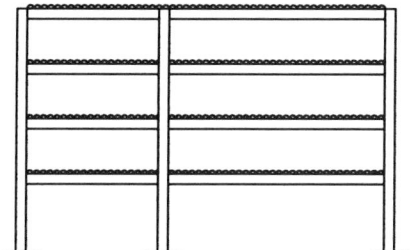

Figure 1.1 *Uneven bay widths*   **Figure 1.2** *Open plan top storey*

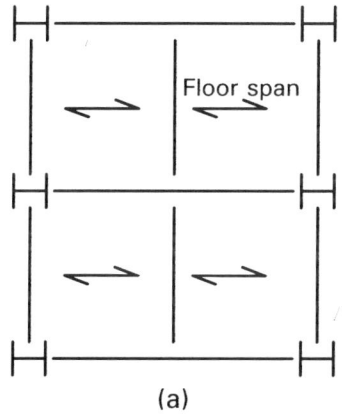   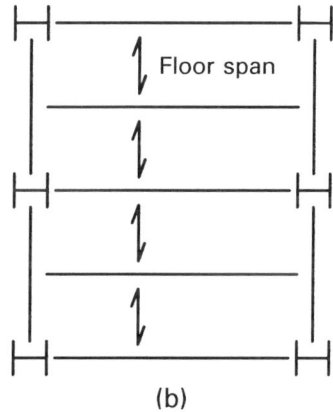

(a)   (b)

Figure 1.3 *Grid of primary and secondary beams with deck spanning (a) parallel and (b) perpendicular to the beam*

1.3.2 Individual Components

For the design procedure to be applicable, the individual components need to comply with the following requirements:

- Frames must have composite beams, in-situ slabs (based on steel decking; see Figure 1.4), and connections. Slabs formed from precast concrete units are excluded because suitable composite connection details for use with this type of construction have not yet been developed, therefore this type of construction was not considered in the background studies.

- Internal connections must be composite flush end plate (non-composite details may not possess sufficient stiffness to ensure frame stability).

- External connections must be either composite flush end plate, bare steel flush end plate or bare steel extended end plate.

- All sections within the frame must be the same steel grade (S275 or S355).

- Universal beams must be used as the steel component of composite beams.

- Universal columns must be used for the columns.

- Universal columns must be at least section size $203 \times 203 \times 60$.

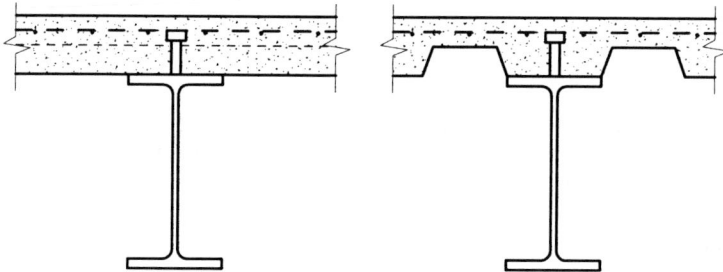

Figure 1.4 *Composite beam sections*

1.3.3 Loading

Loading should be within the limits defined in Table 1.2.

The wind loads may be calculated by using either CP3 Chapt. V[7] or BS 6399-2[8]. The wind loads are taken as concentrated point loads applied at each floor level.

Table 1.2 *Frame loading limits*

	Minimum	Maximum
Dead load on floors (kN/m^2)	3.50	5.00
Imposed load on floors (kN/m^2)	4.00	7.50
Dead load on roof (kN/m^2)	3.75	3.75
Imposed load on roof (kN/m^2)	1.50	1.50
Wind loads (kN)	10	40

1.3.4 Choice of wind-moment method

The designer can rapidly determine whether the wind-moment method will be appropriate for a given frame by using the flowchart provided in Figure 1.5. If the frame is likely to be controlled by SLS sway deflections an alternative design method should be used.

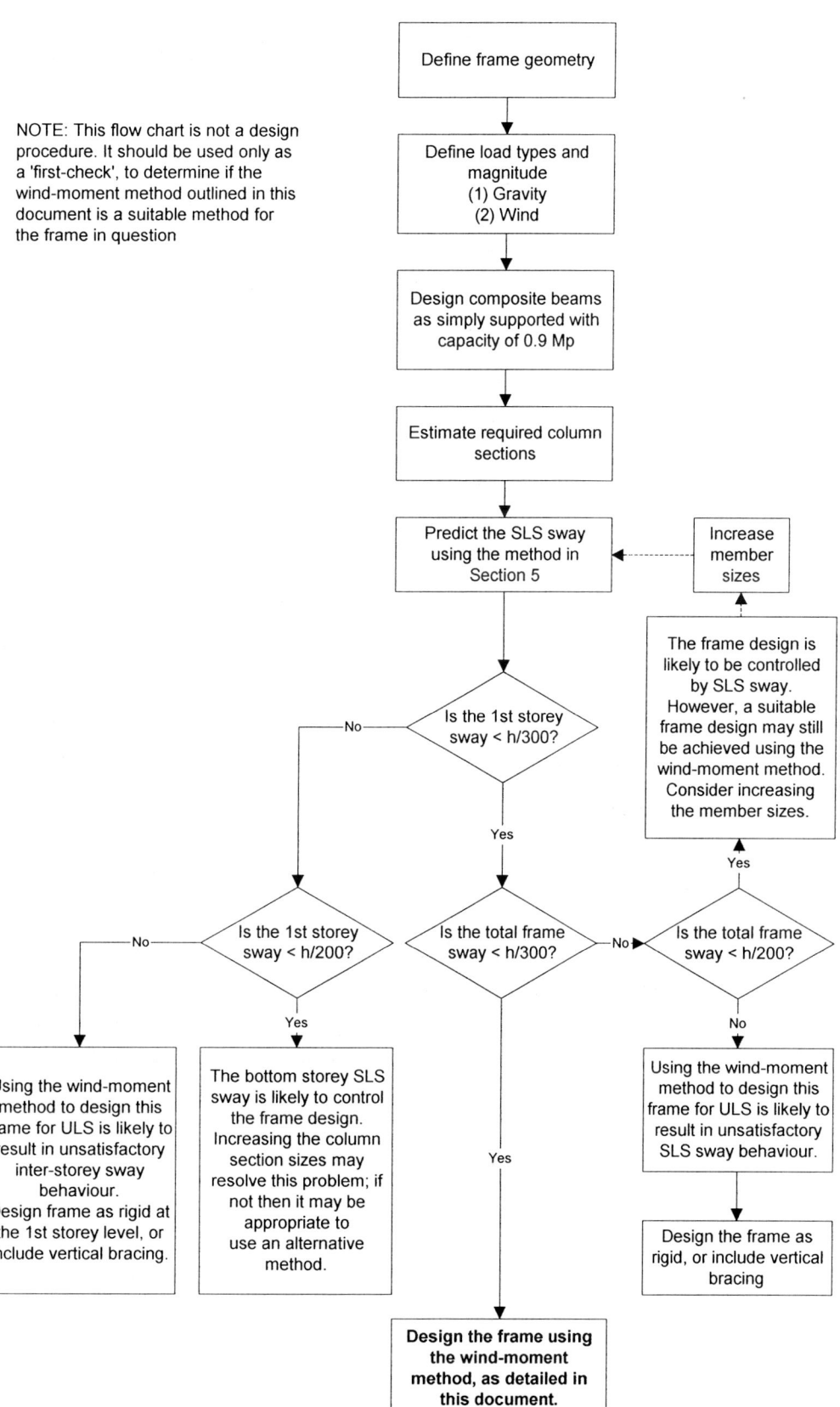

Figure 1.5 *Flow chart to determine the suitability of the wind-moment method for a given frame*

2 PRINCIPLES OF WIND-MOMENT DESIGN

The distinguishing features of the wind-moment method are the basic assumptions that are made at the design stage. These assumptions are that:

- under vertical loads the connections are assumed to act as pins (Figure 2.1(a)),
- under wind loads the connections are assumed to behave as rigid joints, with points of contraflexure occurring at the mid-height of the columns and at the mid-span of the beams (Figure 2.1(b)).

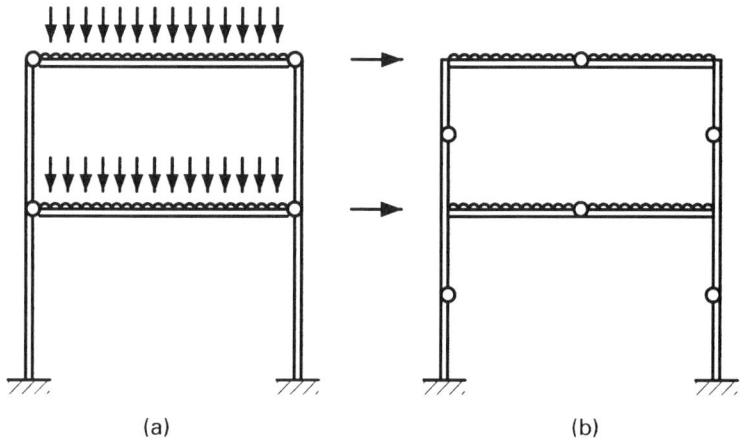

Figure 2.1 *Frame assumptions for the wind-moment method*

2.1 Design method

The first step in the design sequence is to design the composite beams for the ultimate limit state (ULS) fully factored vertical loads, assuming a nominal end fixity moment of 10%. The frame is then analysed under wind loads, with the assumption that the beam-to-column connections behave in a rigid manner. The internal forces and moments are then combined using the principle of superposition, and adopting appropriate load factors for each combination. Design for the ULS is completed by, when necessary, amending the initial section sizes and connection details so that they can withstand the combined effects (Figure 2.2).

Second order, or $P\text{-}\Delta$, effects are accounted for by designing the columns using effective lengths that are greater than the true column lengths. The need for complicated second order calculations is therefore avoided.

The serviceability sway displacements are calculated assuming the beam-to-column connections are rigid. To account for the true flexibility of the connections, a sway amplification factor is used.

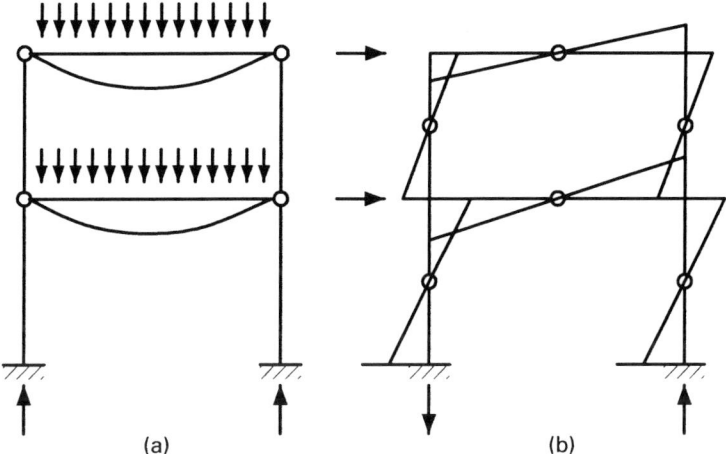

Figure 2.2 *Internal moments and forces according to the wind-moment method*

2.1.1 Beams

Beams designed using the wind-moment method tend to be overdesigned because full account of the true moment resisting nature of the connections is not taken when determining the applied mid-span sagging moments due to vertical loading. Only a nominal 10% reduction in mid-span sagging moment is considered. In reality, substantial hogging moments (up to 40% of the beam sagging moment capacity) may occur at the beam ends, and the subsequent reduction in the sagging moment is not fully exploited in the beam design.

2.1.2 Columns

Columns designed using the wind-moment method tend to be underdesigned due to the adverse effect of the hogging moments developed at the beam ends under vertical loading. This effect is not fully accounted for in the design procedure as only a nominal 10% hogging moment is considered. These hogging moments can be as high as 40% of the composite beam sagging moment capacity, and are particularly significant for external columns or internal columns that are subject to unbalanced loading from adjacent beam spans.

2.1.3 Connections

Unlike bare steel wind-moment connections, composite connections will tend to have excess moment resistance. This is because the area of reinforcement needed to provide the connection with sufficient rotation capacity will often exceed the area needed to provide sufficient moment resistance. The addition of a reinforced concrete slab to a bare steel connection significantly increases the moment capacity of the connection, at the expense of some rotation capacity. However, the most important change to the bare steel connection behaviour, with regard to the wind-moment method, is the increase in connection stiffness; it is this characteristic that provides stability to the frame. The addition of reinforcement substantially increases the bare steel connection stiffness.

2.1.4 Sway displacement

The sway displacements will be larger than those predicted assuming rigid beam-to-column connections. The flexibility of the connections is taken into account using sway amplification factors, as discussed later.

2.1.5 Frame stability

The wind-moment method is, in part, a plastic design method[9]. Accordingly, the following conditions must be satisfied at collapse:

- a mechanism of plastic hinges must have formed
- the internal moments and forces must be in equilibrium with the external loads
- nowhere may the internal moment exceed the plastic moment of resistance.

If the last two conditions are satisfied, the lower-bound theorem states that the applied loads are either less than or equal to the loads that will cause the frame to collapse. These conditions are met by a design using the wind-moment method, which will therefore result in a safe design, provided that the following two criteria are also satisfied:

- the effect of deflections on equilibrium is negligible
- premature collapse does not occur as a result of any form of buckling.

These aspects of the design are satisfied implicitly when plastic or compact sections and ductile connections are used, and the design method given in this publication is adopted (for example with regard to the effective lengths of columns).

Adhering to the recommendations presented in this document will result in a framing system that is capable of resisting the applied loads when the concrete floor slabs are in place, and the concrete has reached the appropriate strength. When the frame is in the bare steel state, the beams and connections will not provide the same level of stability to the frame (because neither the beam nor the connection stiffness is achieved until the slab is structurally effective). However, studies have shown that wind loads on unclad framed structures can be significant during the construction stage of a project[10]. It is important, therefore, that the designer is aware of this, and takes adequate measures to ensure that the structure is stable at all times during the construction phase, as well as when the building is complete. It may therefore be necessary to provide temporary bracing during erection to ensure frame stability.

2.1.6 Foundations

One of the principal frame requirements is the provision of adequate lateral stiffness, so that second order effects remain within acceptable limits. The frame must also conform with acceptable sway displacement limitations under working load conditions. Studies have shown that the degree of rotational restraint provided at the column bases plays a significant part in the overall frame response under lateral loads. Indeed, background studies have shown that the overall sway response varies substantially depending on whether fully encastré bases or nominally rigid bases (based on approximations of base stiffness[11][12][13]) are assumed in the frame analysis. Nominally-pinned bases do not generally provide adequate stability and cannot therefore be used.

Section 3.4 outlines the minimum requirements for base plate details that can be presumed acceptable for frames designed using the wind-moment method.

3 CONNECTIONS

3.1 Connection classification

Connections are classified according to their moment-rotation $(M\text{-}\phi)$ characteristics, with particular regard to strength (moment resistance), stiffness (rigidity), and ductility (rotation capacity).

According to publication *Composite steel-concrete joints in braced frames for buildings*[14], the transition from rigid to semi-rigid connection behaviour for an unbraced frame should be taken as $25EI/L$; where I is the second moment of area of the *uncracked* composite beam. A connection possessing a greater stiffness than this may therefore be classified as 'rigid', even though it does not exhibit infinite stiffness (see Figure 3.1) and will therefore undergo some rotation as moment is applied.

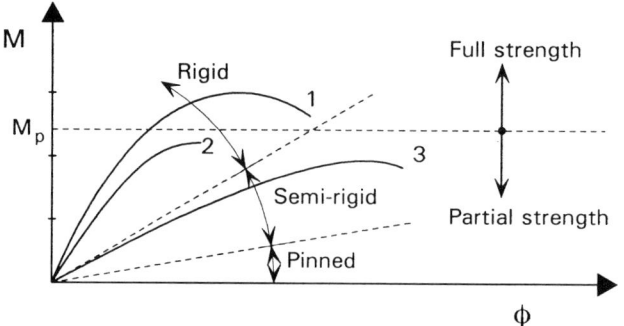

Figure 3.1 *Moment-rotation characteristic*

Connections are described as being full strength, or partial strength, with respect to the adjacent beam. This means that in a composite frame the strength of the connections is defined relative to the negative (hogging) moment resistance of the composite beam.

Curve 1 in Figure 3.1 represents a typical rigid, full strength connection. Curve 2 represents a rigid partial strength connection, and Curve 3 represents a semi-rigid partial strength connection. Composite connections are typically rigid and partial strength (Curve 2) in their final, composite state.

3.2 Beam-to-column connections

The scope of this publication covers only major axis sway frames. Thus, all connections discussed here are to the column flange face.

Only flush end-plate steelwork details should be used as the basis for composite connections. Various forms of stiffening may be necessary to increase the performance of the connections, although it should be noted that increasing the column section size may be a more economical way of improving resistance than adding stiffeners. The need to balance large tensile forces in the reinforcement, across the column, means that it is not unusual to require column web compression stiffeners. Transverse beams that frame into the column web may provide beneficial stiffening to the column web[15].

Design of composite connections is covered by the SCI/BCSA publication *Joints in steel construction: Composite connections*[16]. Included in that publication are design procedures, standard connection details and standard connection capacity tables. Standard connections are described briefly below and Appendix C of this publications presents tables from Reference 16.

3.2.1 Internal connections

Figure 3.2 shows a typical internal connection detail that behaves in a way that is compatible with the wind-moment method of frame design. As explained in Reference 16, the end-plate should be of S275 steel (design grade 43), and not thicker than 20 mm. All bolts should be Grade 8.8 or similar, with one or more tension rows. Standard connection details for this type of connection are given in Appendix C.

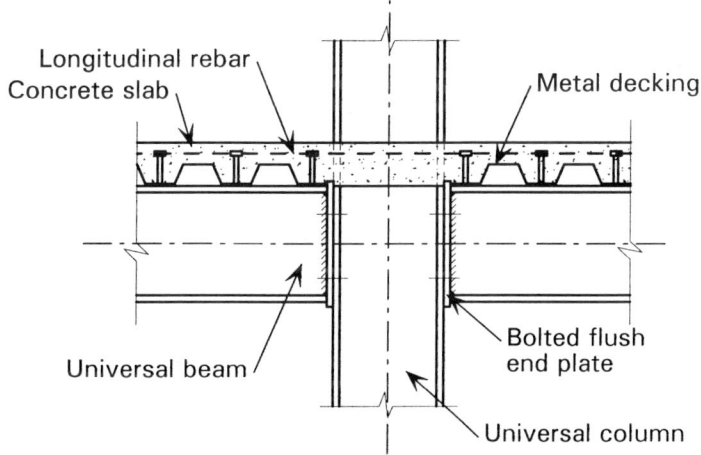

Figure 3.2 *Typical internal connection*

3.2.2 External connections

The frame design is dependent on the type of external connections that are adopted. Figure 3.3 shows the two types of external connection detail that can be used. The detail shown in Figure 3.3(a) includes a small cantilever concrete slab beyond the external column flange. This detailing enables the longitudinal reinforcement to be properly anchored around the column, so that an effective composite connection can be achieved.

The detail shown in Figure 3.3(b) does **not allow** sufficient anchorage of the longitudinal reinforcement, and connections of this type should be designed as bare steel connections in accordance with the SCI/BCSA publication *Joints in steel construction: Moment connections*[17]. Any contribution to moment resistance from the reinforcement is ignored.

Reference 16 recommends the use of non-composite edge connections in braced frames for practical reasons, given that connection rotation stiffness is not required for frame stability. However, for unbraced frames, connection rotation stiffness is required, and hence composite external connections are beneficial.

Regardless of the slab detailing, the steelwork should comply with the recommended bolt and endplate arrangements given in Reference 17. This condition is satisfied by the standard composite connections given in Appendix C.

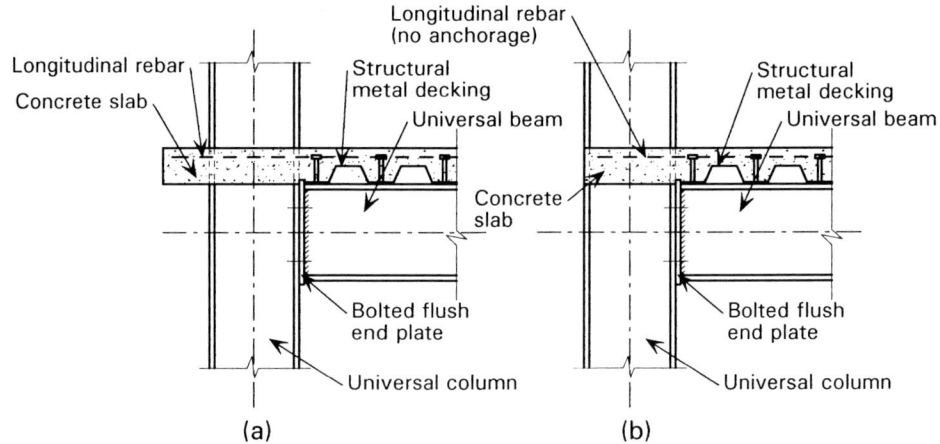

Figure 3.3 *External connection details (a) Composite (b) Effectively Non-composite*

The influence of the external connection detailing on the frame sway displacements is accounted for using the 'sway factors' given in Table 5.1 (Section 5); details as shown in Figure 3.3(b) result in greater sway than when full composite action can be achieved.

When there is insufficient anchorage for the longitudinal reinforcement the external bay may only be considered as actively resisting lateral loading if a bare steel connection of sufficient strength can be adopted.

Connections similar to that shown in Figure 3.3(a) should use end-plates that are no thicker than 20 mm, and bolt locations should be in accordance with the standard layouts presented in Reference 17 and reproduced in Appendix C.

3.2.3 Reinforcement detailing at the connections

The nature of unbraced frames can result in substantial unbalanced moments being generated across internal columns. These moments must be resisted by tension in the longitudinal reinforcement in the slab (at the connection), with the unbalanced tensile forces being resisted by bearing of the concrete slab against the column. The amount of longitudinal reinforcement that can be adopted in the connections must therefore be limited to prevent local crushing of the concrete against the column flange. Transverse reinforcement is also required to prevent splitting of the slab on the low moment side of the connection[14].

Appendix B provides specific details about reinforcement requirements and practical detailing rules. Reinforcing mesh is not considered to act structurally as part of the composite connections, because its limited ductility may lead to failure at relatively low values of connection rotation.

3.3 Standard connections

Moment and shear capacities, as well as detailing guidance, are given for a range of standard composite wind-moment connections in Appendix C. When standard connections are used the designer can be sure that the connection will have the level of rotation capacity and stiffness required by connections for use in wind-moment composite frames designed in accordance with this guide.

The standard connections have the following attributes:

- 12 mm thick flush end plates when M20 bolts are used,
- 15 mm thick flush end plates when M24 bolts are used,
- end plates fabricated from S275 steel,
- full strength flange welds, with a minimum visible fillet of 10 mm,
- continuous 8 mm fillet web welds,
- 1st shear connector at least 100 mm from the column face,
- longitudinal reinforcing bars situated approximately 20 mm above the decking (ie as near to the top of the decking as possible, whilst maintaining sufficient concrete cover).

3.4 Column bases

Columns should be rigidly connected to the foundations in accordance with usual practice for this type of construction. A minimum of four Grade 8.8 bolts should be used, with a minimum base-plate thickness of 25 mm. The centre line of the bolt rows should be at least 50 mm outside of the column flange (measured from the outside face of the column flange), as shown in Figure 3.4. Base plate connections should be checked using the usual design procedures for axial load and moment. More detailed guidance on column bases is given in Reference 17.

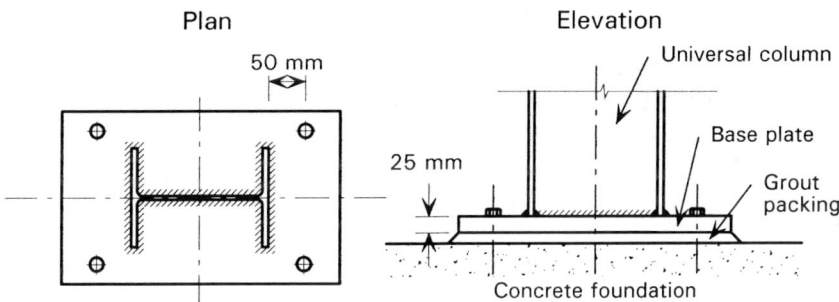

Figure 3.4 *Typical column base-plate connection*

4 DESIGN FOR THE ULTIMATE LIMIT STATE

4.1 Global analysis

4.1.1 Load combinations

The following load combinations should be considered in the design:

1.4 × Dead load + 1.6 × Imposed load + 1.0 × Notional horizontal forces
1.2 × Dead load + 1.2 × Imposed load + 1.2 × Wind load
1.4 × Dead load + 1.4 × Wind load

The notional horizontal forces should be taken as 0.5% of the factored dead plus imposed loads (BS 5950-1:1990 Clauses 5.6.3, 5.1.2.3)[2]

Pattern loading should be considered, in addition to full vertical load applied to all beams.

4.1.2 Internal moments and forces due to vertical loads

Allowance should be made for the end restraint provided by the composite connections at the beam-ends. An end restraint moment equal to 10% of the free bending moment (i.e. the maximum sagging moment for the beam, assuming it to be simply supported) should be considered at the connections (see Figure 4.1).

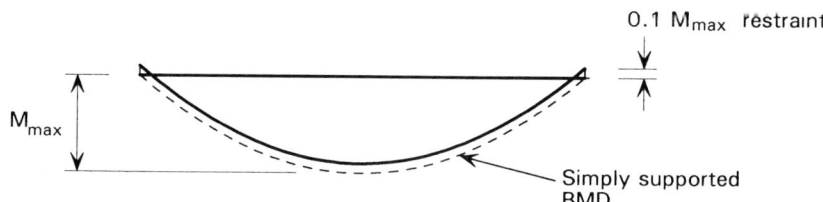

Figure 4.1 *Assumed end restraint moments*

As a result of the vertical loading, each column should be designed to resist the algebraic sum of the restraint moments from opposing beams, plus the moments due to an assumed eccentricity of the beam reactions. The beam reactions are assumed to be applied 100 mm from the column faces (BS 5950-1:1990 Clause 4.7.6). The net moment applied to a column at any given level should be divided between the column lengths above and below that level in accordance with BS 5950-1:1990 Clause 4.7.7. It is assumed that this moment has no effect at other levels.

4.1.3 Internal moments and forces due to horizontal loads

Analysis should be by the 'portal method', and based on the following assumptions:

- horizontal loads are applied at floor levels
- there is a point of contraflexure at the mid-height of each column
- there is a point of contraflexure at the mid-span of each beam

- each bay acts as a simple portal, and the total horizontal load applied to the frame is divided between the bays in proportion to their spans.

Algebraic formulae based on these assumptions are given in Appendix A.

4.2 Design of beams

4.2.1 Construction stage

The bare steel beam sections must be capable of resisting the construction stage loading (BS 5950-3: Section 3.1 Clause 2.2.3)[3]. The construction loads should be treated as imposed loads, using appropriate load factors from BS 5950-1[2].

The lateral restraint offered to the steel beams, and therefore their moment capacity will depend on the direction in which the steel decking spans. When the ribs of the decking run parallel to the beam (Figure 1.3(a)) no lateral restraint is provided, and the effective length of the beam should be taken as equal to the distance between any secondary (transverse) beams. Conversely, decking spanning perpendicularly to the beam (Figure 1.3(b)) does provide lateral restraint when it is appropriately fixed to the beam.

4.2.2 Moment resistance (sagging)

Composite beams should be Class 1, plastic, or Class 2, compact, and within the scope detailed in Section 1.2.2. The plastic moment resistance of the composite beam can be determined from the formulae presented in BS 5950-3: Section 3.1: Appendix B1[3] or in Reference 4.

Beams should be designed as simply supported, with the sagging moment resistance limited to 90% of the plastic moment resistance.

4.2.3 Moment resistance (hogging)

The hogging resistance of the composite beam may be determined using formulae presented in BS 5950-3:Section 3.1 Appendix B.2.4. Only the bare steel section and the longitudinal reinforcing bars (not mesh) are considered when determining the hogging resistance of the composite beam; cracked concrete is assumed to have zero strength.

Detailing must be such that the hogging resistance of the composite beam is greater than the moment resistance of the adjacent (partial strength) connection.

4.2.4 Shear connection

The degree of shear connection provided should be consistent with the limits given in BS 5950-3:Section 3.1 Clause 5.4.5.5 and Clause 5.5.2. Full shear connection should be provided in hogging moment regions.

When determining the required number of shear studs (which are 25% weaker in hogging regions than in sagging), the hogging length at each end of the beam should be taken as span/5.

4.3 Design of columns

4.3.1 Effective lengths

When determining the compression resistance for a major axis sway frame where the minor axis is braced, the values of P_{cx} and P_{cy} should be based on the following effective lengths:

- For in-plane behaviour (bending about the major axis):

  $L_{Ex} = 1.5 L$

- For out-of-plane behaviour (bending about the minor axis):

  $L_{Ey} = 1.0 L$

Values of P_{cx} and P_{cy} are tabulated in the SCI publication *Steelwork design guide to BS 5950: Part 1: 1990 - Volume 1 Section properties and member capacities*[18].

4.3.2 Equivalent slenderness for buckling resistance moment

When determining the buckling resistance moment (BS 5950-1:1990 Clauses 4.3.7.3 and 4.3.7.4), the slenderness λ_{LT} should be taken as $\lambda_{LT} = 0.5\ (L/r_y)$. Values of M_{bs} are tabulated in Reference 18.

4.3.3 Design moments

For each load combination (Section 4.1.1) the column moments should be taken as the sum of:

- the net (i.e. out of balance) moments due to the assumed eccentricity of the beam reactions arising from vertical load (BS 5950-1:1990 Clause 4.7.6),
- the net moments due to the assumed '10%' restraint moments at the beam ends arising from vertical loads (BS 5950-1:1990 Clause 2.1.2.4),
- the moments due to horizontal loads (i.e. wind or notional forces).

Because the horizontal loads may reverse, the design moments should be determined by addition of the numerical magnitudes of each of the component moments.

4.3.4 Class of section

Sections should be Class 1 (plastic) or Class 2 (compact), so that they can attain their plastic moment resistance M_p.

4.3.5 Overall buckling check

The following relationship (BS 5950-1:1990, Clause 4.8.3.3) should be satisfied:

$$\frac{F_c}{P_c} + \frac{M_x}{M_{bs}} + \frac{M_y}{p_y Z_y} \leq 1.0 \qquad (1)$$

where:

F_c is the applied compression load in the member

P_c is the compression resistance (taken as the lesser of P_{cy} and P_{cx} calculated for L_{Ey} and L_{Ex} respectively)

M_x is the applied moment about the major axis

M_{bs} is the buckling resistance moment for simple design

M_y is the applied moment about the minor axis

p_y is the design strength of the steel

Z_y is the elastic section modulus about the minor axis.

4.4 Design of connections

4.4.1 Applied moments and forces

For each load combination (refer to Section 4.1.1), the moments at the connections should be taken as the sum of:

- the '10%' restraint moments due to partial fixity at the beam ends under vertical load (refer to Section 4.1.2)
- the moments due to horizontal loads (i.e. wind or notional forces).

The vertical shear forces at the connections should be taken as the sum of:

- the beam end shears due to vertical load
- the shear forces in the beams due to horizontal loading on the frame.

The connections should be designed for both maximum (hogging) and minimum (potentially sagging) moments because the direction of the horizontal loads may reverse. However, if sagging does occur it is acceptable for the connections to be understrength in sagging, provided that this understrength is compensated by a corresponding overstrength in hogging. The design criterion is not then that the hogging strength of the connection exceeds the maximum applied hogging moment, and that the sagging strength exceeds the maximum applied sagging moment, but rather that the sum of the hogging and sagging resistances exceeds the absolute sum of the applied hogging and sagging moments. The resistance of a composite connection in sagging may be conservatively taken as that of the bare steel detail (with the lower bolts in tension). For a more typical case where sagging does not occur (there is simply a reduction in hogging as the wind direction reverses) the connection should be designed for the larger hogging moment, and nominal resistance to sagging will suffice.

4.4.2 Design procedure

As mentioned in Section 3.3, it is strongly recommended that standard connections are used. The connection design procedure then requires only a simple comparison of the maximum absolute sum of the applied hogging and sagging moments with the tabulated connection strengths in order to identify a suitable standard detail. Resistances and detailing guidance for standard connection details are tabulated in Appendix C. Additional guidance on reinforcement detailing is given in Appendix B.

Reference should be made to *Joints in steel construction - Composite connections*[16] for full design procedures and principles relating to composite connections. Although the connection design procedures given in that publication are for simplicity described as being applicable to braced frames only, they are appropriate for use in wind-moment frames which comply with the scope specified in Section 1.2 of this publication. This 'relaxation' is possible because for such frames the connections will not in practice be subject to significant sagging moments.

In addition to possessing adequate strength, the connections must possess sufficient rotation capacity to behave as a plastic hinge. This can be achieved by appropriate detailing. The standard connections presented in this guide satisfy this criterion for both propped and unpropped construction.

5 DESIGN FOR SERVICEABILITY LIMIT STATE

5.1 General

Sway behaviour, and in particular the prediction of sway displacements, is very complex. The magnitude of sway displacements is influenced by a number of factors: relative member stiffnesses, connection characteristics, the ratio of horizontal to vertical loading, and column base behaviour. All these factors have a significant influence on the sway response of an unbraced frame.

It has been found that composite frames designed using the ULS procedures given in this guide will, in most practical cases, conform to the common sway limit of $h/300$. Indeed, for frames with two or more 9 m or 12 m span bays, the sway displacement is likely to be well below this limit even under high wind loads. However, for some frames that are subjected to relatively severe wind loading, the sway displacements within the first storey may be critical. This is particularly true for 'slender' two bay frames. Conversely, numerical studies[5] have demonstrated that if the first storey sway of a frame is less than about $h/200$, then the frame will generally prove adequate under ultimate limit state conditions.

It should be remembered that the common sway limit of $h/300$[2][19] is not a strict performance criterion; rather the limit is intended to be compared with calculated deflections for unclad frames, recognising that the beneficial influence of cladding and other components is not accounted for in such calculations. Although the limit of $h/300$ appears to have proved successful in preventing damage to cladding and glazing systems in existing buildings by preventing excessive displacements, its justification rests solely on this 'track record'. Alternative limits may be acceptable in some cases.

It has been found that accounting for the true flexibility of the beam-to-column connections in a composite wind-moment frame increases the sway displacements by about 30% compared with a similar frame analysed assuming fully rigid connections[5]. This increase is allowed for in wind-moment design by using the sway amplification factors that are presented below.

5.2 Sway prediction

5.2.1 Initial analysis

A frame which has been designed for the ultimate limit state should, as a first step, be analysed as an elastic rigidly-jointed frame to determine the sway displacements. Unfortunately this procedure is complicated by the fact that, unlike bare steel beams, the cross-sectional properties of composite beams are not constant along their length. Composite beam properties vary significantly depending on the type and magnitude of loading applied, and the strength and stiffness of the restraining end connections because these factors effect the length of beam in hogging. It is not sufficiently accurate to consider the composite beam to have a constant flexural stiffness (equal to that of the uncracked cross-section) along its entire length when attempting to calculate sway displacements of a composite frame. However, using an equivalent beam model (as described in

Section 5.2.2) with a simple graphical method of sway prediction (as described in Section 5.2.3) excellent results can be obtained.

5.2.2 Equivalent beam model

An expression for *equivalent* beam stiffness has been developed[5] based on the expected moment distribution along the length of the beam (see Figure 5.1). The equivalent beam stiffness is given as:

$$I_{eq} = \frac{7.5\, I_g\, I_n^2}{9\, I_n^2 + 2\, I_g\, I_n} \tag{2}$$

where:

I_g is the second moment of area of the uncracked 'sagging' composite beam

I_n is the second moment of area of the cracked 'hogging' composite beam.

Procedures for calculating I_g and I_n are given in BS 5950-3:Section 3.1:1990 Clause B.3.

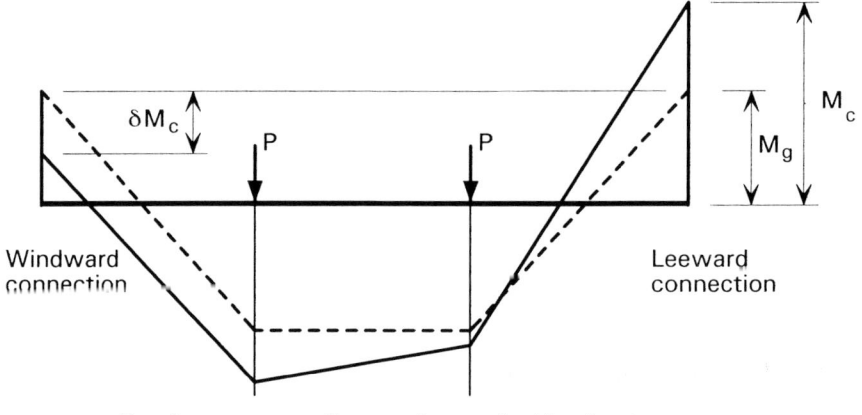

---- = Bending moment diagram for vertical load only
——— = Bending moment diagram for vertical load and wind load

Figure 5.1 *Schematic bending moment diagram for a beam in a sway frame*

5.2.3 Graphical method of sway prediction

Studies[5] have shown that sway predictions, based on the graphical procedure detailed in a paper by Wood and Roberts[20], and modified using the sway amplification factors given in Table 5.1, are in excellent agreement with the results obtained using more sophisticated analysis. The sway amplification factors are used to allow for the 'wind-moment connections' being more flexible than 'rigid' details.

Table 5.1 *Sway amplification factors*

External connection type	Frame sway amplification factor
Composite (Figure 3.3(a))	1.4
Non-composite (Figure 3.3(b))	1.6

5.2.4 Sway limits

In addition to checking that the overall frame sway deflection is less than $h/300$, it is important to check that the sway within each storey is less than $h/300$ (where h is the storey height)[2]. The bottom storey sway is likely to be the most critical; its magnitude can be estimated from the total frame sway using the values given in Table 5.2.

Table 5.2   *Bottom storey sway percentages*

Total number of storeys	Bottom storey sway as a percentage of total frame sway
2 storeys	80%
3 storeys	65%
4 storeys	55%

5.2.5 Redesign for stiffness

If sway deflections are unacceptable, the frame may be modified to increase its stiffness as an alternative to redesigning as a braced frame. Member sizes could be increased, remembering that increases in the beam depth, or the column flange thickness, will also increase the connection stiffness (although they may not necessarily increase its moment resistance).

6 REFERENCES

1. SALTER, P.R., COUCHMAN, G.H. and ANDERSON, D.
   Wind-moment design of low rise frames (P263)
   The Steel Construction Institute, 2000

2. BRITISH STANDARDS INSTITUTION
   BS 5950: Structural use of steelwork in building
   BS 5950-1:1990: Code of practice for design in simple and continuous construction

3. BRITISH STANDARDS INSTITUTION
   BS 5950: Structural use of steelwork in building
   BS 5950-3: Design in composite construction
   Section 3.1:1990: Code of practice for design of simple and continuous composite beams

4. LAWSON, R.M.
   Design of composite slabs and beams with steel decking (P055)
   The Steel Construction Institute, 1989

5. HENSMAN, J.S.
   Investigation of the wind-moment method for unbraced composite frames.
   M.Phil thesis, University of Nottingham, 1998

6. ANDERSON, D., READING, S.J. and KAVIANPOUR, K.
   Wind-moment design for unbraced frames (P082)
   The Steel Construction Institute, 1991

7. BRITISH STANDARDS INSTITUTION
   CP 3: Basic data for the design of buildings
   Chapter V: Loading
   Part 2: Wind load
   BSI, 1972

8. BRITISH STANDARDS INSTITUTION
   BS 6399: Loading for buildings
   BS 6399-2:1997: Code of practice for wind loads

9. NEAL, B.G.
   Plastic Methods of Structural Analysis
   Chapman & Hall Ltd and Science Paperbacks, 1970

10. WILLFORD, M.R. and ALSOP, A.C.
    Design guide for wind loads on unclad building structures during construction
    Building Research Establishment, 1990

11. GUISSE, S., VANDEGANS, D. and JASPART, J-P.
Application of the component method to column bases - experimentation and development of a mechanical model for characterization. Report No. MT 195
Research Centre of the Belgian Metalworking Industry, Steel Construction Dept., Dec. 1996

12. CIMsteel: EUREKA PROJECT 130
Modelling of steel structures for computer analysis (P148)
The Steel Construction Institute, 1995

13. WALD, F., BAUDUFFE, N., SOKOL, Z. and MUZEAU J-P.
Pre-design model of the column-base stiffness
IABSE Colloquium, Istanbul, 1996

14. COST, C.I.
Composite steel-concrete joints in braced frames for buildings
ed. Anderson, D.
European Commission, Brussels, 1996

15. LI, T.Q., NETHERCOT, D.A. and CHOO, B.S.
Behaviour of flush end-plate composite connections with unbalanced moment and variable shear/moment ratios - I Experimental behaviour
Journal of Constructional Steel Research, pp 125-164, Vol. 38 (2), 1996

16. THE STEEL CONSTRUCTION INSTITUTE and THE BRITISH CONSTRUCTIONAL STEELWORK ASSOCIATION
Joints in Steel Construction: Composite Connections (P213)
SCI/BCSA, 1998

17. THE STEEL CONSTRUCTION INSTITUTE and THE BRITISH CONSTRUCTIONAL STEELWORK ASSOCIATION
Joints in Steel Construction: Moment Connections (P207)
SCI/BCSA, 1995

18. THE STEEL CONSTRUCTION INSTITUTE
Steelwork Design Guide to BS 5950: Part 1: 1990
Volume 1: Section properties and member capacities (P202)
SCI, 1997

19. BRITISH STANDARDS INSTITUTION
Eurocode 3: Design of Steel Structures
DD ENV 1993-1-1:1992: General rules and rules for buildings (including UK NAD)

20. WOOD, R.H. and ROBERTS, E.H.
A graphical method of predicting sidesway in multistorey buildings
Proceedings of the Institution of Civil Engineers, Part 2, Vol. 59, pp 353-272, June 1975

21. BRITISH STANDARDS INSTITUTION
BS 8110: Structural use of concrete
BS 8110-1:1985: Code of practice for design and construction

22. BRITISH STANDARDS INSTITUTION
BS 4449:1988: Specification for carbon steel bars for the reinforcement of concrete

23. BRITISH STANDARDS INSTITUTE
DD ENV 10080:1996: Steel for the reinforcement of concrete weldable ribbed reinforcing steel B500. Technical delivery conditions for bars, coils and welded fabric (including UK NAD)

24. BRITISH STANDARDS INSTITUTION
BS 4466:1989: Specification for scheduling, dimensioning, bending and cutting of reinforcement for concrete

25. LI, T.Q., NETHERCOT, D.A. and CHOO, B.S.
Behaviour of flush end plate composite connections with unbalanced moment and variable shear/moment ratios - II prediction of moment capacity.
Journal of Constructional Steel Research, Vol 38 (2), 1996

26. LEON, R.T.
Composite connections
Progress in Structural Engineering and Materials Vol 1(2): pp159-169
1998

APPENDIX A: Portal method of analysis

The following guidance is reproduced from the publication *Wind-moment design of low rise frames*[1].

A.1 Introduction

The forces and moments in a multi-storey, multi-bay wind-moment frame can be determined by simple manual calculation using the so-called portal method. The wind and notional horizontal forces are shared between the bays according to the relative bay widths, and the forces in the beams and columns are calculated for this distribution of loading. A detailed explanation is given below for part of a multi-storey two-bay frame.

A.2 Distribution of horizontal load

Each bay is assumed to act as a single portal and the total horizontal load is divided between the bays in proportion to their spans. For a two bay frame, the loads in the two separate bays (as shown in Figure A.1b) are given by:

$$H_{1,1} = L_1 W_1 /(L_1 + L_2); \quad H_{2,1} = L_2 W_1 /(L_1 + L_2)$$
$$H_{1,2} = L_1 W_2 /(L_1 + L_2); \quad H_{2,2} = L_2 W_2 /(L_1 + L_2)$$
(A.1)

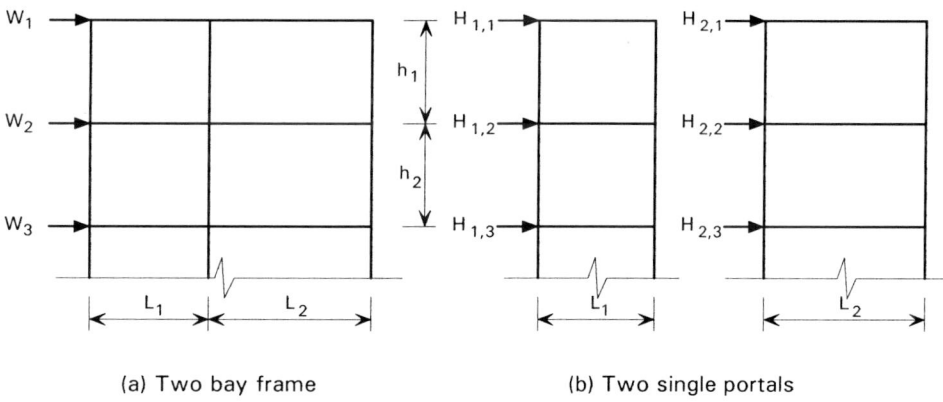

(a) Two bay frame (b) Two single portals

Figure A.1 *Distribution of horizontal load*

A.3 Calculation of internal forces in columns

The forces acting on a part of one bay and the pin locations assumed in wind-moment design are shown in Figure A.2a.

The forces acting on the portion of the bay above the points of contraflexure at A and D are shown in Figure A.2b. The horizontal force H_1 is assumed to be divided equally between the two columns. Thus

$$S_1 = H_1/2$$
(A.2)

The vertical forces F_1 can be found by taking moments about the point of contraflexure at either A or D:

$$F_1 L = H_1 h_1 / 2$$

which gives:

$$F_1 = H_1 h_1 / (2L) \tag{A.3}$$

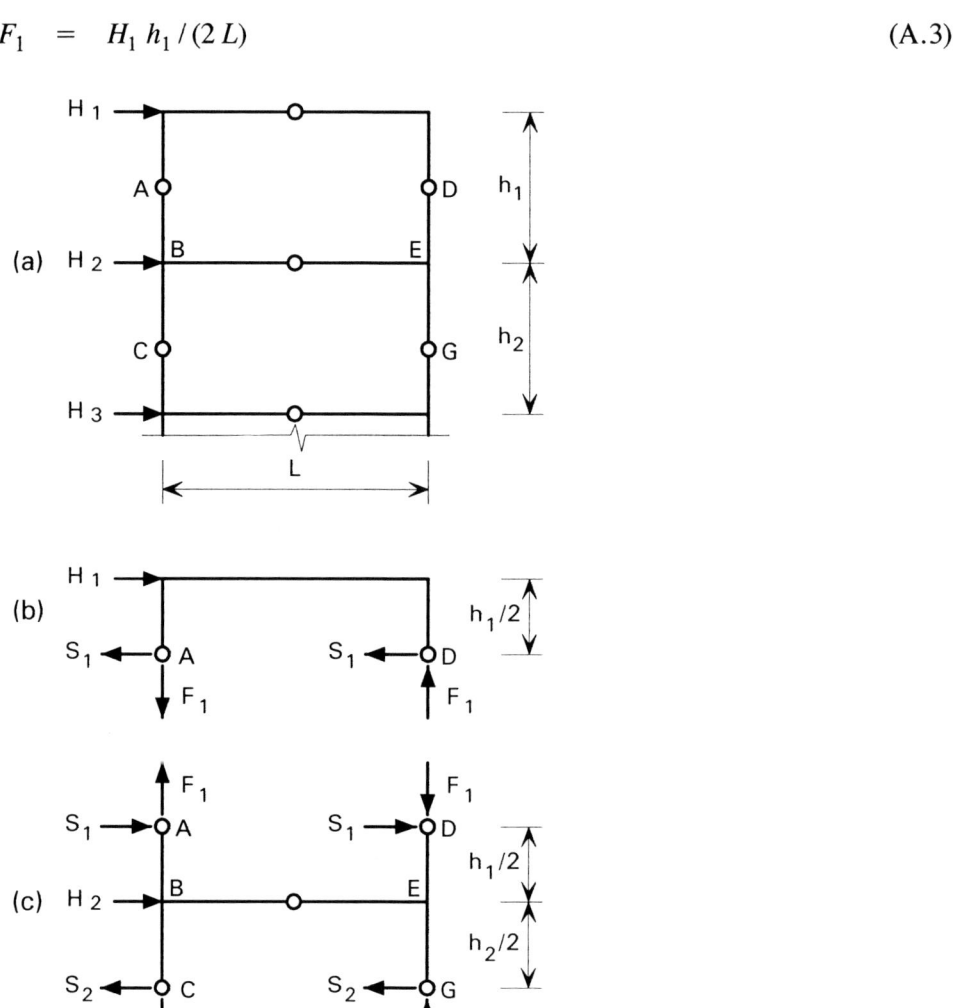

Figure A.2 *Internal forces in columns*

The forces acting on the portion ABCDEG of the bay are shown in Figure A.2c. It follows from the assumption above that:

$$S_2 = (H_1 + H_2)/2 \tag{A.4}$$

Taking moments about the point of contraflexure at either C or G:

$$F_2 L = H_2 h_2 / 2 + 2 S_1 (h_1 + h_2)/2 + F_1 L$$

Substituting for S_1 and F_1 and re-arranging:

$$F_2 = H_1 h_1 / L + (H_1 + H_2) h_2 / (2L) \tag{A.5}$$

A.4 Calculation of internal moments

It is clear from Figure A.2b that the internal moment at the head of each column is given by:

$M_1 = S_1 h_1 / 2$

Substituting for S_1:

$M_1 = H_1 h_1 / 4$ (A.6)

For equilibrium, the moment at each end of the roof beam is also equal to M_1. The bending moment diagram is shown in Figure A.3a.

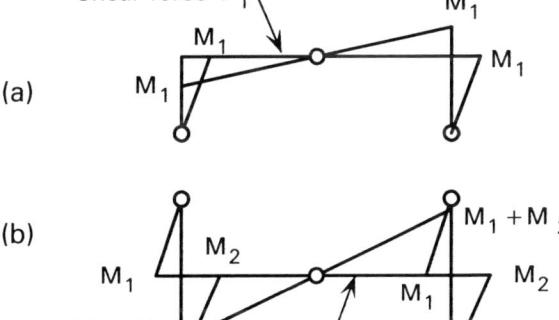

Figure A.3 *Internal moments*

Referring to Figure A.2c, the internal moment in each upper column at B and E is also M_1. The corresponding moment in the lower columns is given by:

$M_2 = S_2 h_2 / 2$

Substituting for S_2:

$M_2 = (H_1 + H_2) h_2 / 4$ (A.7)

For equilibrium at B and E, the internal moment at each end of the beam BE equals $(M_1 + M_2)$, as shown in Figure A.3b.

A.5 Calculation of shear forces in beams

As a point of contraflexure is assumed at the mid-length of each beam (Figure A.3), the shear force in the roof beam is given by:

$V_1 = M_1 / (L/2)$

Substituting for M_1:

$V_1 = H_1 h_1 / (2L)$ (A.8)

Similarly, the shear force V_2 in beam BE is given by:

$V_2 = (M_1 + M_2)/(L/2)$

Substituting for M_1 and M_2:

$V_2 = H_1(h_1 + h_2)/(2L) + H_2 h_2/(2L)$ (A.9)

A.6 Forces and moments in an internal column

These are obtained by summing the values calculated for adjacent bays on either side of the column.

It is found that the vertical forces in an internal column due to horizontal loading are zero.

APPENDIX B: Connection detailing

The following information is based on that found in *Joints in steel construction: Composite connections*[16], Sections 5 and 4.2. References to compact section capacity have been omitted because plastic hinge analysis is invoked in wind-moment frame design.

B.1 Introduction

Appropriate connection detailing is necessary in order to:

- prevent premature failure of the tension bolts or reinforcement
- ensure sufficient deformation takes place to generate the tension bolt and reinforcement forces assumed in the design
- prevent concrete crushing against the column under unbalanced loading.
- ensure that the connections have sufficient rotation capacity to form a plastic hinge.

The detailing rules that follow should be used in conjunction with the minimum reinforcement area requirements given in Table B.1, and the maximum area limitations presented in Section B2.2.

B.1.1 Reinforcement and shear connection

Conventional reinforcement detailing according to BS 8110[21] should be adopted. Bar diameters should not be less than 16 mm, since smaller diameter bars are generally less ductile. Effective anchorage of the reinforcement is achieved by curtailing the bars in the compression zone of the slab. This zone normally starts at about 0.2 times the beam span on either side of the support, and sufficient anchorage length should be provided beyond this point, as in conventional reinforced concrete practice[21] (for example 40 times the bar diameter for a 'Type 2 deformed' bar in concrete with a cube strength of 30 N/mm^2). To ensure adequate bar anchorage for composite connections to perimeter columns, the longitudinal bars should be 'wrapped' around the columns as shown schematically in Figure B.1 b). The cantilever dimension, c, for perimeter connections will depend on the reinforcement detailing to ensure adequate cover and anchorage. Although reinforcing mesh may also be present in the slab to control cracking, its contribution to moment capacity should be ignored.

The following detailing rules are shown schematically in Figure B.1. Limitations on the positions of reinforcing bars ensure that they can work effectively as components in a truss to resist unbalanced loads.

1. Longitudinal reinforcing bars should be uniformly spaced either side of the column, with the nearest bars approximately 20 mm from the column edge, to achieve adequate cover. The furthest bars to be included in the effective area should not be more than approximately $2b_c$ from the column centreline (dimension e_L).
   NOTE: e_L *is a function of the column width, rather than the beam span.*

2. Transverse reinforcing bars (which are necessary to resist forces in the concrete 'behind' the column when unbalanced loading is applied) should not

extend more than the required anchorage length[21]) beyond points $2b_c$ either side of the column centre line. This should ensure that the orthogonal connection behaviour is not significantly influenced by these bars.

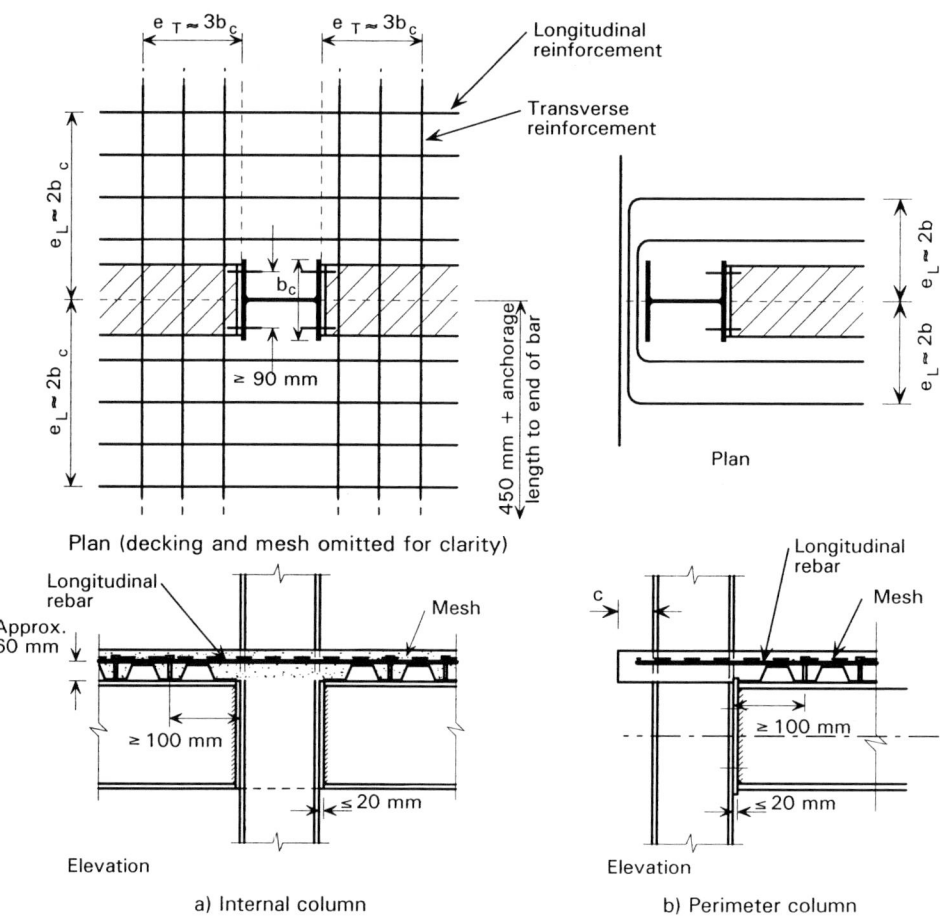

Figure B.1 *Geometrical detailing rules - beam-to-column composite connections*

3. Transverse reinforcing bars should be uniformly spaced either side of the column (Figure B.1), with the nearest bars approximately 20 mm from the column edge. The furthest bars included in the effective area should not be more than approximately $3b_c$ from the column face (dimension e_T).

4. Longitudinal reinforcing bars should be placed approximately 20 mm above the top of the decking, to ensure adequate concrete cover. The position of the transverse bars above or below the longitudinal bars will depend on the orientation of the decking and the depth of slab above the decking. All the bars must be positioned so that they have adequate cover to the decking and the top of slab[21]).

Keeping the longitudinal bars close to the top of the decking minimises the strain they must undergo to achieve a given rotation.

5. The first shear connector should be at least 100 mm from the face of the column.

This limitation ensures that reinforcing bars are strained over a substantial length, so that sufficient rotation can take place.

B.1.2 Steelwork

1. The end plate thickness should be not more than 60% of the bolt diameter; 12 mm for M20 bolts and 15 mm for M24 bolts. End plates should be made from S275 steel.

2. Horizontal spacing of the bolts (gauge) should be not less than 90 mm.

These steelwork restrictions ensure that, as the connection rotates, end plate deformation is the 'weak link'. Steelwork details complying with these rules have been shown in tests to possess sufficient rotation capacity.

B.2 Reinforcement area limits

B.2.1 Minimum area of reinforcement

In general, the rotation capacity of a connection increases as the area of reinforcement increases[16]. A minimum area is therefore needed to ensure sufficient rotation capacity for a plastic connection. Minimum areas that should be provided are given in Table B.1 as a function of:

- beam size
- beam steel grade
- reinforcement properties

Table B.1 *Minimum area of reinforcement - 'plastic' connections*

Steel	Rebar elongation limit	Beam depth (mm)							
		203	254	305	356	406	457	533	610
S275	5%	500	500	500	650	1100	1450	1800	3000
	10%	500	500	500	500	500	600	750	1150
S355	5%	500	500	600	1400	2100	3100	-	-
	10%	500	500	500	500	650	900	2000	2850

Note: A dash (-) in the table indicates that excessive reinforcement is required.

The minimum values in Table B.1 marked '5%' should normally be used. These values are appropriate for connections using high yield bars, grade 460B complying with British Standard BS 4449[22]. This grade of reinforcement has a mandatory requirement of 14% minimum elongation at fracture, and a non-mandatory requirement of 5% minimum elongation at maximum force. Grade B500B bars complying with BS EN 10080[23] are required to have similar properties; they must be able to achieve 5% **total** elongation at maximum force. Elongation at fracture and elongation at maximum force are illustrated in Figure B.2.

Minimum reinforcement areas are also given in Table B.1 for connections that use reinforcement which is capable of achieving 10% total elongation at maximum force. The increased reinforcement ductility offers considerable advantages in some cases because it permits the use of less reinforcement.

It is essential that when a design is based on the use of 10% elongation bars, this is made clear in the project specification. This can be done by giving the bars an 'X' designation, rather than the 'T' generally used for high tensile bars[24]. The 'X' informs the contractor that the bars need specific, non-standard properties. It is recommended that, if possible, the reinforcement supplier uses coloured labels to clearly distinguish the high elongation 'X' bars on site. In case of doubt concerning the elongation capacity of bars, approximately half the UK manufacturers provide reinforcement suppliers with appropriate test information. It should therefore be relatively easy for the contractor to confirm suitability with his reinforcement supplier.

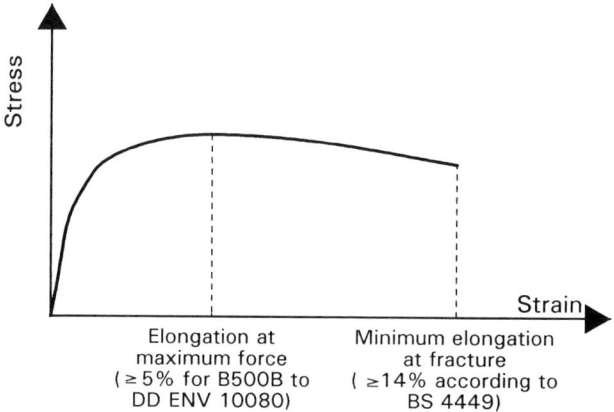

Figure B.2 *Elongation limits for reinforcement*

Bars that are currently produced using a hot forming process may be assumed to be appropriate for use with the '10%' limits. All 20 mm diameter bars produced by major manufacturers in the UK currently are hot formed, as are, often, 16 mm bars.

B.2.2 Maximum area of reinforcement

The reinforcement area must also be limited to a maximum value in order to:

- prevent local concrete crushing failure under unbalanced loading
- keep the compression zone in the lower half of the steel beam.

The reasons for these limits are discussed below.

In order to consider the potential concrete crushing failure, a truss model has been developed to represent how double sided composite connections behave when the applied moments on either side are unequal[14]. Figure B.3 illustrates the components in the truss, showing that the connection resistance relies on the ability of the concrete to bear against the column on the low moment side. The net force in the reinforcement is therefore limited by the strength and area of concrete in bearing. An enhancement factor may be applied to the concrete strength because of its confinement[15][25].

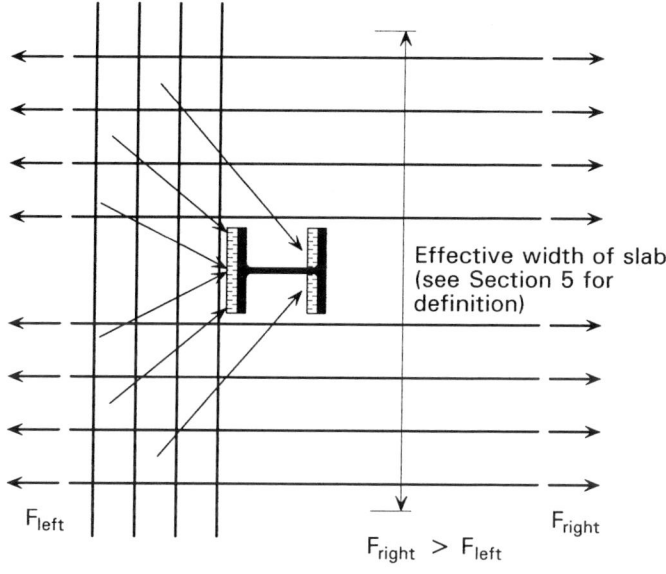

Effective truss members formed by:
• longitudinal reinforcement
• transverse reinforcement
• concrete

Figure B.3 *Truss model for connection behaviour under unbalanced moment*

According to the truss model, the area of longitudinal reinforcement must not exceed:

$$A_L \leq \frac{1.5\, b_c\, d_s\, f_{cu}}{\mu\, f_y} \tag{B.1}$$

where:

1.5   is a factor that includes allowances for the enhanced concrete strength due to confinement[15][25], the total area of concrete in bearing[26] and the material partial safety factor

   Note: This factor has been modified from the value given in Reference 16 as a result of the application of more recent research[15][25][26].

b_c is the width of column

d_s is the depth of slab above decking

f_{cu} is the cube strength of concrete

f_y is the yield strength of the rebar

μ is a function of the difference in applied moments, and beam depths, either side of the node, given by:

$$\mu = 1 - \frac{M_{low}}{M_{high}} \frac{h_{r1}}{h_{r2}} \qquad (B.2)$$

where:

M_{low} is the smaller applied moment (may be taken as the larger connection moment resistance less the vector addition of the applied moments. The vector addition is the difference in the applied moments if both are hogging, and the sum if one is sagging)

M_{high} is the larger applied moment (may be taken as the larger connection moment resistance)

h_{r1} is the reinforcement lever arm on the high moment side

h_{r2} is the reinforcement lever arm on the low moment side.

Transverse reinforcement acts as a tension member in the truss model (see Figure B.3). The area of transverse reinforcement must satisfy the following limit:

$$A_T \geq \frac{0.35 \, \mu \, A_L}{\left(\dfrac{e_T}{e_L} - 0.3 \right)} \qquad (B.3)$$

where:

$e_L \approx 2.0 h_c$ (this is the outer limit of the longitudinal reinforcement from the column centre line)

$e_T \approx 3.0 b_c$

e_L and e_T are identified in Figure B.1.

It is assumed that the longitudinal and transverse reinforcing bars have the same nominal yield strength.

In theory, the length of transverse reinforcement must be limited so that, whilst sufficient anchorage is provided for the bars to act in the truss, they do not affect the behaviour of the 'transverse beam connections'. In practice, this should not be critical. Detailing rules are given in Section B1.1.

Additional reasons for imposing an upper limit on the reinforcement area are:

- to ensure adequate strain in the reinforcement, compression must be restricted to the lower half of the steel beam (i.e. the plastic neutral axis must not be higher than the mid-depth of the web)

- to avoid the need for column compression stiffeners. The standard connection capacity tables, in Appendix C, indicate whether or not a compression stiffener is required in the column.

APPENDIX C: Capacity Tables

The information given in this appendix is a representation of the data from Appendix B of *Joints in steel construction: composite connections*[16]. A full explanation of the calculation of the moment capacity is given in that publication.

Composite Connections

	1 ROW M20 8.8 BOLTS 200 x 12 S275 END PLATE	BEAM SIDE

| BEAM
Serial
Size | Effective reinforcement (option, number and size of bars, A_{reinf}, F_{reinf}) ||||||||||||||||
| | A
4 No. φ16
804 mm²
351 kN || B
6 No. φ16
1210 mm²
529 kN || C
8 No. φ16
1610 mm²
704 kN || D
10 No. φ16
2010 mm²
878 kN || E
4 No. φ20
1260 mm²
551 kN || F
6 No. φ20
1890 mm²
826 kN || G
8 No. φ20
2510 mm²
1097 kN || H
10 No. φ20
3140 mm²
1372 kN ||
	'A' mm	M_C kNm	'A' mm	M_C kNm	'A' mm	M_C kNm	'A' mm	M_C kNm	'A' mm	M_C kNm	'A' mm	M_C kNm	'A' mm	M_C kNm	'A' mm	M_C kNm
457x191x98	<u>398</u>	<u>268*</u>	398	362*	398	454*	398	546*	398	373*	398	518*	398	*661**	398	807
89	<u>395</u>	<u>266*</u>	395	359*	395	451*	395	543*	395	371*	395	515*	395	*648**	395	802
82	<u>392</u>	<u>265*</u>	392	358*	392	449*	392	540*	392	369*	392	513*	392	*654**	—	—
74	<u>390</u>	<u>264*</u>	390	356*	390	447*	384	*531**	390	367*	*390*	*510**	*390*	*651**	—	—
67	<u>387</u>	<u>262*</u>	387	354*	*387*	*444**	387	*535**	387	365*	*387*	*508**	—	—	—	—
457x152x82	<u>396</u>	<u>267*</u>	396	361*	396	453*	396	545*	396	372*	396	517*	396	*660**	—	—
74	<u>394</u>	<u>266*</u>	394	359*	394	450*	*382*	*529**	394	370*	*386*	*506**	*387*	*648**	—	—
67	<u>391</u>	<u>264*</u>	391	356*	*385*	*443**	391	*538**	391	368*	391	*511**	—	—	—	—
60	<u>388</u>	<u>263*</u>	388	355*	388	*445**	—	—	388	366*	—	—	—	—	—	—
52	<u>384</u>	<u>261*</u>	384	352*	—	—	—	—	384	363*	—	—	—	—	—	—
406x178x74	345	239*	345	323*	345	406*	*345*	*489*	345	333*	345	464*	345	*593*	—	—
67	342	237*	342	321*	*342*	*403**	*336*	*486*	342	331*	*342*	*461**	—	—	—	—
60	340	236*	340	319*	340	*401**	—	—	340	330*	*340*	*459**	—	—	—	—
54	337	234*	*337*	*317**	*337*	*399**	—	—	*333*	*324**	—	—	—	—	—	—
406x140x46	338	235*	*338*	*317**	—	—	—	—	*338*	*328**	—	—	—	—	—	—
39	*334*	*232**	—	—	—	—	—	—	—	—	—	—	—	—	—	—
356x171x67	296	211*	296	286*	296	*361*	*289*	*428*	296	296*	296	*413*	*296*	*528*	—	—
57	292	209*	292	284*	*292*	*357*	—	—	292	293*	*292*	*409*	—	—	—	—
51	289	207*	*289*	*282**	*289*	*355*	—	—	289	*291**	—	—	—	—	—	—
45	287	206*	*287*	*280**	—	—	—	—	287	*289**	—	—	—	—	—	—
356x127x39	*281*	*203**	—	—	—	—	—	—	—	—	—	—	—	—	—	—
33	*280*	*202**	—	—	—	—	—	—	—	—	—	—	—	—	—	—
305x165x54	244	182	244	248	244	*313*	—	—	244	256	*244*	*359*	—	—	—	—
46	241	180	*241*	*246*	—	—	—	—	*238*	*254*	—	—	—	—	—	—
40	*238*	*179*	*238*	*244*	—	—	—	—	*238*	*252*	—	—	—	—	—	—
305x127x48	244	182	*237*	*239*	*239*	*308*	—	—	*235*	*244*	—	—	—	—	—	—
42	241	181	*224*	*221*	—	—	—	—	*222*	*225*	—	—	—	—	—	—
37	*232*	*171*	*231*	*233*	—	—	—	—	*229*	*238*	—	—	—	—	—	—
305x102x33	*230*	*163*	—	—	—	—	—	—	—	—	—	—	—	—	—	—
28	*211*	*143*	—	—	—	—	—	—	—	—	—	—	—	—	—	—
25	—	—	—	—	—	—	—	—	—	—	—	—	—	—	—	—
254x146x43	193	151	*188*	*195*	—	—	—	—	*187*	*199*	—	—	—	—	—	—
37	*191*	*149*	*175*	*176*	—	—	—	—	*172*	*179*	—	—	—	—	—	—
31	*177*	*128*	—	—	—	—	—	—	—	—	—	—	—	—	—	—
254x102x28	*172*	*121*	—	—	—	—	—	—	—	—	—	—	—	—	—	—
25	—	—	—	—	—	—	—	—	—	—	—	—	—	—	—	—

398   Beam may be either grade S275 or grade S355
369   Beam must be grade S355 to satisfy neutral axis position requirements
<u>264</u>   Beam must be grade S275 to satisfy minimum reinforcement requirements (see Table B.1)
*   Reinforcement requires a guaranteed strain at maximum load of at least 10% for S355 beams, and possibly for S275 beams (check using Table B.1)
256   Connection capacity exceeds 0.8 M_p of composite beam in hogging (see ref. 16 Section 3.2.1 for significance of this)
The value of F_{r1} is based on the assumption that the NA is at least 200 mm below the bolt row. It should be reduced in accordance with ref. 16 Section 4.2 Step 1D when necessary.

Composite Connections

1 ROW M20 8.8 BOLTS
200 x 12 S275 END PLATE

COLUMN SIDE

S275											Column Serial Size	S355										
Panel Shear Cap.	Web Compn. Cap.	Tension Zone	Compression Zone									Compression Zone								Tension Zone	Web Compn. Cap.	Panel Shear Cap.
		F_{r1}	Reinforcement option									Reinforcement option								F_{r1}		
(kN)	(kN)	(kN)	A	B	C	D	E	F	G	H		A	B	C	D	E	F	G	H	(kN)	(kN)	(kN)
											356x368											
1000	1141	✓	✓	✓	✓	✓	✓	✓	S	S	x 202	✓	✓	✓	✓	✓	✓	✓	S	✓	1486	1302
849	935	✓	✓	✓	✓	S	✓	S	S	S	x 177	✓	✓	✓	✓	✓	✓	S	S	✓	1217	1105
725	766	✓	✓	✓	S	S	✓	S	S	S	x 153	✓	✓	✓	S	✓	S	S	S	✓	974	944
605	605	✓	✓	S	S	S	S	S	S	S	x 129	✓	✓	S	S	✓	S	S	S	✓	788	787
											305x305											
1037	1432	✓	✓	✓	✓	✓	✓	✓	✓	S	x198	✓	✓	✓	✓	✓	✓	✓	✓	✓	1865	1350
816	1051	✓	✓	✓	✓	S	✓	✓	S	S	x158	✓	✓	✓	✓	✓	✓	✓	S	✓	1368	1062
703	858	✓	✓	✓	S	S	✓	S	S	S	x137	✓	✓	✓	✓	✓	✓	S	S	✓	1116	915
595	692	✓	✓	S	S	S	S	S	S	S	x118	✓	✓	S	S	✓	S	S	S	✓	909	774
503	553	✓	S	S	S	S	S	S	S	S	x97	✓	✓	S	S	S	S	S	S	✓	713	649
											254x254											
882	1384	✓	✓	✓	✓	✓	✓	✓	✓	S	x167	✓	✓	✓	✓	✓	✓	✓	✓	✓	1802	1149
685	992	✓	✓	✓	✓	S	✓	S	S	S	x132	✓	✓	✓	✓	✓	✓	S	S	✓	1292	892
551	744	✓	✓	✓	S	S	S	S	S	S	x107	✓	✓	✓	S	✓	S	S	S	✓	969	717
434	557	✓	S	S	S	S	S	S	S	S	x89	✓	S	S	S	S	S	S	S	✓	725	566
360	436	✓	S	S	S	S	S	S	S	S	x73	✓	S	S	S	S	S	S	S	✓	563	465
											203x203											
459	701	✓	✓	S	S	S	S	S	S	S	x86	✓	✓	✓	S	✓	S	S	S	✓	913	598
353	512	✓	S	S	S	S	S	S	S	S	x71	✓	S	S	S	S	S	S	S	✓	666	460
322	440	✓	S	S	S	S	S	S	S	S	x60	✓	S	S	S	S	S	S	S	✓	568	415

Tension Zone:
✓ Column satisfactory for bolt row tension values shown for the beam side.
195 Recalculate moment capacity based on reduced bolt row force (195 kN) using dimension 'A' to derive appropriate lever arm - or provide tension stiffener at the appropriate bolt row level.

Compression Zone:
✓ Column capacity exceeds $\Sigma F = F_{reinf} + F_{r1}$
S Provide compression stiffener.

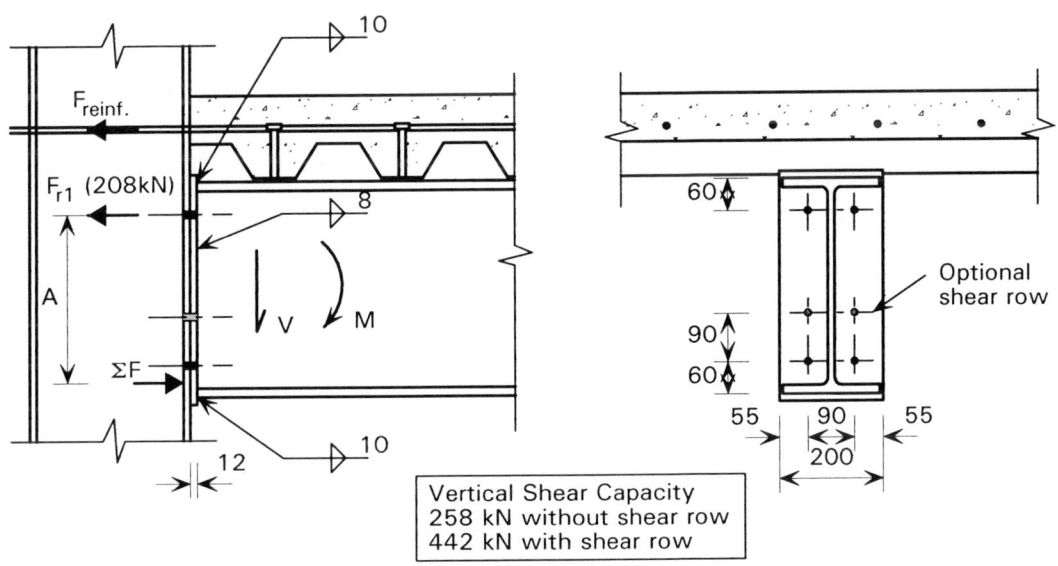

Vertical Shear Capacity
258 kN without shear row
442 kN with shear row

Composite Connections

	2 ROWS M20 8.8 BOLTS 200 x 12 S275 END PLATE	**BEAM SIDE**

	Effective reinforcement (option, number and size of bars, A_{reinf}, F_{reinf})														
BEAM Serial Size	A 4 No. φ16 804 mm² 351 kN		B 6 No. φ16 1210 mm² 529 kN		C 8 No. φ16 1610 mm² 704 kN		D 10 No. φ16 2010 mm² 878 kN		E 4 No. φ20 1260 mm² 551 kN		F 6 No. φ20 1890 mm² 826 kN		G 8 No. φ20 2510 mm² 1097 kN		H 10 No. φ20 3140 mm² 1372 kN	
	'A' mm	M_C kNm	'A' mm	M_C kNm	'A' mm	M_C kNm	'A' mm	M_C kNm	'A' mm	M_C kNm	'A' mm	M_C kNm	'A' mm	M_C kNm	'A' mm	M_C kNm
533x210x122	<u>384</u>	<u>363*</u>	<u>384</u>	<u>470*</u>	384	575*	384	681*	<u>384</u>	<u>483*</u>	384	649*	384	813*	377	967*
109	<u>380</u>	<u>360*</u>	<u>380</u>	<u>466*</u>	380	571*	380	676*	<u>380</u>	<u>480*</u>	380	645*	380	807*	**380**	**973***
101	<u>378</u>	<u>358*</u>	<u>378</u>	<u>465*</u>	378	569*	378	674*	<u>378</u>	<u>478*</u>	378	642*	378	*791***	378	969*
92	<u>375</u>	<u>357*</u>	<u>375</u>	<u>462*</u>	375	566*	375	*670***	<u>375</u>	<u>475*</u>	375	639*	375	800*	—	—
82	<u>372</u>	<u>354*</u>	<u>372</u>	<u>459*</u>	372	563*	**372**	**666***	<u>372</u>	<u>472*</u>	*361*	*623***	—	—	—	—
457x191x98	<u>308</u>	<u>310*</u>	308	403*	308	495*	308	588*	308	415*	308	560*	*301*	*693***	**308**	**848**
89	<u>305</u>	<u>307*</u>	305	401*	305	492*	*305*	*584***	305	412*	305	556*	305	699*	—	—
82	<u>302</u>	<u>306*</u>	302	398*	302	490*	*302*	*572***	302	410*	*302*	*553***	302	695*	—	—
74	<u>300</u>	<u>304*</u>	300	396*	*300*	*487***	300	578*	300	408*	**300**	**551***	—	—	—	—
67	<u>297</u>	<u>302*</u>	297	394*	**297**	**484***	—	—	297	405*	297	548*	—	—	—	—
457x152x82	<u>306</u>	<u>309*</u>	306	402*	306	494*	294	571*	306	414*	297	548*	*301*	*693***	—	—
74	<u>304</u>	<u>307*</u>	304	400*	*295*	*482***	304	583*	304	411*	285	531*	—	—	—	—
67	<u>301</u>	<u>305*</u>	301	397*	285	472*	**293**	**570***	296	404*	**295**	**546***	—	—	—	—
60	<u>298</u>	<u>303*</u>	*288*	*386***	292	479*	—	—	286	395*	—	—	—	—	—	—
52	<u>287</u>	<u>295*</u>	**289**	**387***	—	—	—	—	**287**	**397***	—	—	—	—	—	—
406x178x74	255	273	255	357*	255	440*	243	502	255	368*	*247*	*485***	250	618	—	—
67	252	271*	252	355*	*244*	*425***	**252**	**520**	252	365*	**252**	**495***	—	—	—	—
60	250	270*	*250*	*353***	250	435*	—	—	245	357*	—	—	—	—	—	—
54	247	268*	*233*	*331***	—	—	—	—	**247**	**361***	—	—	—	—	—	—
406x140x46	*237*	*255***	—	—	—	—	—	—	—	—	—	—	—	—	—	—
30	**235**	**255***	—	—	—	—	—	—	—	—	—	—	—	—	—	—

398   Beam may be either grade S275 or grade S355
369   Beam must be grade S355 to satisfy neutral axis position requirements
<u>264</u>   Beam must be grade S275 to satisfy minimum reinforcement requirements (see Table B.1)
*   Reinforcement requires a guaranteed strain at maximum load of at least 10% for S355 beams, and possibly for S275 beams (check using Table B.1)
256   Connection capacity exceeds 0.8 M_p of composite beam in hogging (see ref. 16 Section 3.2.1 for significance of this)
The value of F_{r1} is based on the assumption that the NA is at least 200 mm below the bolt row. It should be reduced in accordance with ref. 16 Section 4.2 Step 1D when necessary.

Composite Connections

| 2 ROWS M20 8.8 BOLTS 200 x 12 S275 END PLATE | COLUMN SIDE |

S275											Column Serial Size	S355										
Panel Shear Cap.	Web Compn. Cap.	Tension Zone		Compression Zone									Compression Zone								Tension Zone		Web Compn. Cap.	Panel Shear Cap.
		F_{r1}	F_{r2}	Reinforcement option									Reinforcement option								F_{r1}	F_{r2}		
(kN)	(kN)	(kN)		A	B	C	D	E	F	G	H		A	B	C	D	E	F	G	H	(kN)		(kN)	(kN)
												356x368												
1000	1141	✓	✓	✓	✓	✓	S	✓	S	S	S	x 202	✓	✓	✓	✓	✓	✓	✓	S	✓	✓	1486	1302
849	935	✓	✓	✓	✓	S	S	✓	S	S	S	x 177	✓	✓	✓	S	✓	✓	S	S	✓	✓	1217	1105
725	766	✓	✓	✓	S	S	S	S	S	S	S	x 153	✓	✓	S	S	✓	S	S	S	✓	✓	974	944
605	605	✓	✓	S	S	S	S	S	S	S	S	x 129	✓	S	S	S	S	S	S	S	✓	✓	788	787
												305x305												
1037	1432	✓	✓	✓	✓	✓	✓	✓	✓	S	S	x198	✓	✓	✓	✓	✓	✓	✓	✓	✓	✓	1865	1350
816	1051	✓	✓	✓	✓	✓	S	✓	S	S	S	x158	✓	✓	✓	✓	✓	S	S	S	✓	✓	1368	1062
703	858	✓	✓	✓	S	S	S	S	S	S	S	x137	✓	✓	✓	S	✓	S	S	S	✓	✓	1116	915
595	692	✓	✓	S	S	S	S	S	S	S	S	x118	✓	✓	S	S	✓	S	S	S	✓	✓	909	774
503	553	✓	✓	S	S	S	S	S	S	S	S	x97	✓	S	S	S	S	S	S	S	✓	✓	713	649
												254x254												
882	1384	✓	✓	✓	✓	✓	✓	✓	✓	S	S	x167	✓	✓	✓	✓	✓	✓	✓	✓	✓	✓	1802	1149
685	992	✓	✓	✓	✓	S	S	✓	S	S	S	x132	✓	✓	✓	✓	✓	S	S	S	✓	✓	1292	892
551	744	✓	✓	✓	S	S	S	S	S	S	S	x107	✓	✓	S	S	✓	S	S	S	✓	✓	696	717
434	557	✓	✓	S	S	S	S	S	S	S	S	x89	✓	S	S	S	S	S	S	S	✓	✓	725	566
360	436	✓	✓	S	S	S	S	S	S	S	S	x73	S	S	S	S	S	S	S	S	✓	✓	563	465
												203x203												
459	701	✓	✓	✓	S	S	S	S	S	S	S	x86	✓	✓	S	S	✓	S	S	S	✓	✓	913	598
353	512	✓	✓	S	S	S	S	S	S	S	S	x71	S	S	S	S	S	S	S	S	✓	✓	666	460
322	440	✓	✓	S	S	S	S	S	S	S	S	x60	S	S	S	S	S	S	S	S	✓	✓	568	415

Tension Zone:
✓   Column satisfactory for bolt row tension values shown for the beam side.
195   Recalculate moment capacity based on reduced bolt row force (195 kN) using dimension 'A' to derive appropriate lever arm - or provide tension stiffener at the appropriate bolt row level.

Compression Zone:
✓   Column capacity exceeds $\Sigma F = F_{reinf} + F_{r1}$
S   Provide compression stiffener.

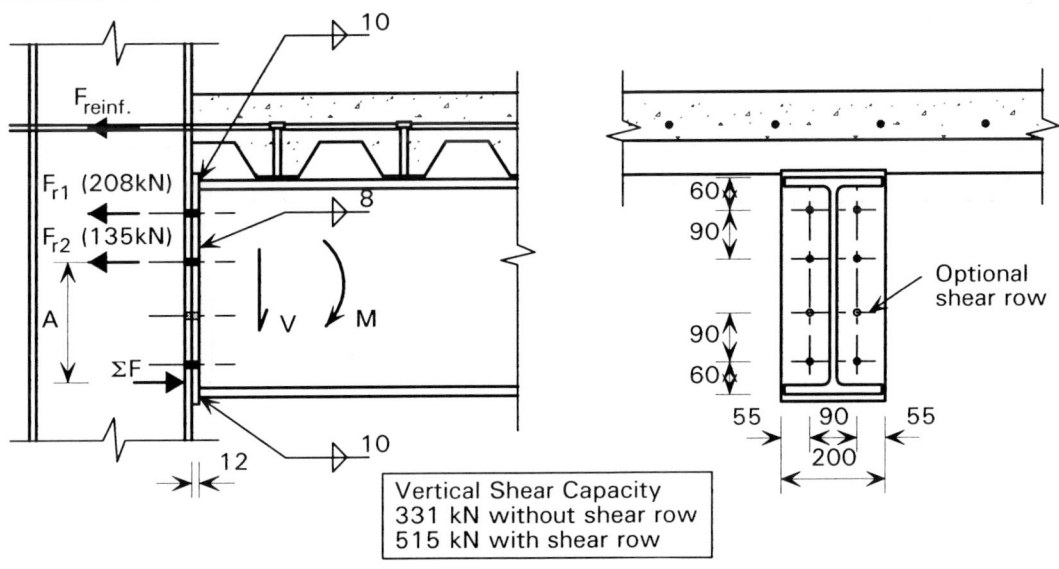

Vertical Shear Capacity
331 kN without shear row
515 kN with shear row

Composite Connections

	2 ROWS M20 8.8 BOLTS 250 x 12 S275 END PLATE	BEAM SIDE

BEAM Serial Size	Effective reinforcement (option, number and size of bars, A_{reinf}, F_{reinf})															
	A 4 No. φ16 804 mm² 351 kN		B 6 No. φ16 1210 mm² 529 kN		C 8 No. φ16 1610 mm² 704 kN		D 10 No. φ16 2010 mm² 878 kN		E 4 No. φ20 1260 mm² 551 kN		F 6 No. φ20 1890 mm² 826 kN		G 8 No. φ20 2510 mm² 1097 kN		H 10 No. φ20 3140 mm² 1372 kN	
	'A' mm	M_c kNm	'A' mm	M_c kNm	'A' mm	M_c kNm	'A' mm	M_c kNm	'A' mm	M_c kNm	'A' mm	M_c kNm	'A' mm	M_c kNm	'A' mm	M_c kNm
533x210x122	384	375*	384	482*	384	588*	384	693*	384	495*	384	662*	384	825*	*375*	*976*
109	380	372*	380	479*	380	583*	380	688*	380	492*	380	657*	375	811*	380	985*
101	378	371*	378	477*	378	581*	378	686*	378	490*	378	654*	*367*	*800*	378	981*
92	375	369*	375	474*	375	578*	375	682*	375	487*	375	651*	*375*	*812*	—	—
82	372	366*	372	471*	366	569*	372	678*	372	484*	359	631*	—	—	—	—
457x191x98	308	319*	308	413*	308	505*	308	597*	308	425*	308	570*	*299*	*700*	308	858
89	305	317*	305	410*	305	502*	*305*	*594*	305	422*	305	566*	**305**	**708***	—	—
82	305	315*	302	408*	302	499*	293	579*	302	419*	296	556*	**302**	**705***	—	—
74	300	314*	300	406*	294	491*	*300*	*588*	300	417*	300	560*	—	—	—	—
67	297	312*	297	404*	297	494*	—	—	297	415*	293	552*	—	—	—	—
457x152x82	306	319*	306	412*	306	504*	292	578*	306	423*	295	555*	**300**	**701***	—	—
74	304	316*	304	409*	293	489*	*304*	*592*	304	421*	283	534*	—	—	—	—
67	301	314*	296	402*	283	475*	*291*	*577*	294	412*	294	553*	—	—	—	—
60	298	312*	285	393*	—	—	—	—	298	416*	—	—	—	—	—	—
52	284	303*	287	394*	—	—	—	—	285	404*	—	—	—	—	—	—

398   Beam may be either grade S275 or grade S355
369   Beam must be grade S355 to satisfy neutral axis position requirements
264   Beam must be grade S275 to satisfy minimum reinforcement requirements (see Table B.1)
*   Reinforcement requires a guaranteed strain at maximum load of at least 10% for S355 beams, and possibly for S275 beams (check using Table B.1)
256   Connection capacity exceeds 0.8 M_p of composite beam in hogging (see ref. 16 Section 3.2.1 for significance of this)
The value of F_{r1} is based on the assumption that the NA is at least 200 mm below the bolt row. It should be reduced in accordance with ref. 16 Section 4.2 Step 1D when necessary.

Composite Connections

2 ROWS M20 8.8 BOLTS
250 x 12 S275 END PLATE
COLUMN SIDE

S275											Column Serial Size	S355										
Panel Shear Cap.	Web Compn. Cap.	Tension Zone		Compression Zone								Compression Zone								Tension Zone		Web Compn. Cap.	Panel Shear Cap.
		F_{r1}	F_{r2}	Reinforcement option								Reinforcement option								F_{r1}	F_{r2}		
(kN)	(kN)	(kN)		A	B	C	D	E	F	G	H		A	B	C	D	E	F	G	H	(kN)		(kN)	(kN)
1000	1141	✓	✓	✓	✓	✓	S	✓	S	S	S	356x368 x 202	✓	✓	✓	✓	✓	✓	✓	S	✓	✓	1486	1302
849	935	✓	✓	✓	✓	S	S	✓	S	S	S	x 177	✓	✓	✓	S	✓	✓	S	S	✓	✓	1217	1105
725	766	✓	✓	✓	S	S	S	S	S	S	S	x 153	✓	✓	S	S	✓	S	S	S	✓	✓	974	944
605	605	✓	✓	S	S	S	S	S	S	S	S	x 129	✓	S	S	S	S	S	S	S	✓	✓	788	787
1037	1432	✓	✓	✓	✓	✓	✓	✓	✓	S	S	305x305 x198	✓	✓	✓	✓	✓	✓	✓	✓	✓	✓	1865	1350
816	1051	✓	✓	✓	✓	S	S	✓	S	S	S	x158	✓	✓	✓	✓	✓	✓	S	S	✓	✓	1368	1062
703	858	✓	✓	✓	S	S	S	✓	S	S	S	x137	✓	✓	✓	S	✓	S	S	S	✓	✓	1116	915
595	692	✓	✓	S	S	S	S	S	S	S	S	x118	✓	✓	S	S	S	S	S	S	✓	✓	909	774
503	553	✓	✓	S	S	S	S	S	S	S	S	x97	S	S	S	S	S	S	S	S	✓	✓	713	649
882	1384	✓	✓	✓	✓	✓	✓	✓	✓	S	S	254x254 x167	✓	✓	✓	✓	✓	✓	✓	✓	✓	✓	1802	1149
685	992	✓	✓	✓	✓	S	S	✓	S	S	S	x132	✓	✓	✓	✓	✓	✓	S	S	✓	✓	1292	892
551	744	✓	✓	✓	S	S	S	✓	S	S	S	x107	✓	✓	S	S	✓	S	S	S	✓	✓	969	717
434	557	✓	✓	S	S	S	S	S	S	S	S	x89	S	S	S	S	S	S	S	S	✓	✓	725	566
360	436	✓	✓	S	S	S	S	S	S	S	S	x73	S	S	S	S	S	S	S	S	✓	✓	563	465
459	701	✓	✓	S	S	S	S	S	S	S	S	203x203 x86	✓	✓	S	S	S	S	S	S	✓	✓	913	598
353	512	✓	✓	S	S	S	S	S	S	S	S	x71	S	S	S	S	S	S	S	S	✓	✓	666	460
322	440	✓	✓	S	S	S	S	S	S	S	S	x60	S	S	S	S	S	S	S	S	✓	✓	568	415

Tension Zone:
✓ Column satisfactory for bolt row tension values shown for the beam side.
195 Recalculate moment capacity based on reduced bolt row force (195 kN) using dimension 'A' to derive appropriate lever arm - or provide tension stiffener at the appropriate bolt row level.

Compression Zone:
✓ Column capacity exceeds $\Sigma F = F_{reinf} + F_{r1}$
S Provide compression stiffener.

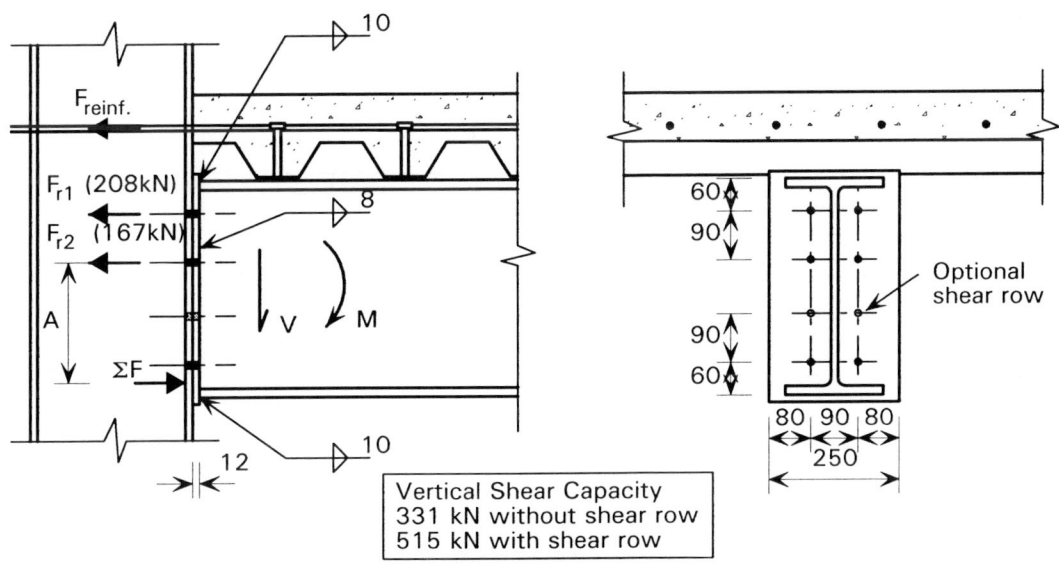

Vertical Shear Capacity
331 kN without shear row
515 kN with shear row

Composite Connections

1 ROW M24 8.8 BOLTS
200 x 15 S275 END PLATE

BEAM SIDE

BEAM Serial Size	Effective reinforcement (option, number and size of bars, A_{reinf}, F_{reinf})															
	A 4 No. φ16 804 mm² 351 kN		B 6 No. φ16 1210 mm² 529 kN		C 8 No. φ16 1610 mm² 704 kN		D 10 No. φ16 2010 mm² 878 kN		E 4 No. φ20 1260 mm² 551 kN		F 6 No. φ20 1890 mm² 826 kN		G 8 No. φ20 2510 mm² 1097 kN		H 10 No. φ20 3140 mm² 1372 kN	
	'A' mm	M_C kNm	'A' mm	M_C kNm	'A' mm	M_C kNm	'A' mm	M_C kNm	'A' mm	M_C kNm	'A' mm	M_C kNm	'A' mm	M_C kNm	'A' mm	M_C kNm
457x191x98	398	307*	398	401*	398	493*	398	585*	398	412*	398	557*	392	693*	398	846
89	395	305*	395	398*	395	490*	395	581*	395	410*	395	554*	395	696*	—	—
82	392	303*	392	396*	392	487*	387	572*	392	407*	392	551*	392	693*	—	—
74	390	302*	390	394*	390	485*	390	576*	390	405*	390	549*	—	—	—	—
67	387	300*	387	392*	387	482*	—	—	387	403*	387	545*	—	—	—	—
457x152x82	396	306*	396	400*	396	492*	386	572*	396	411*	389	548*	396	699*	—	—
74	394	304*	394	397*	387	482*	394	580*	394	409*	378	536*	—	—	—	—
67	391	302*	391	395*	378	473*	385	570*	391	406*	391	549*	—	—	—	—
60	388	301*	381	386*	388	483*	—	—	379	396*	—	—	—	—	—	—
52	380	295*	381	386*	—	—	—	—	379	397*	—	—	—	—	—	—
406x178x74	345	272*	345	357*	345	440*	336	512	345	367*	339	491*	345	626	—	—
67	342	271*	342	354*	336	431*	342	520	342	365*	342	495*	—	—	—	—
60	340	269*	340	353*	340	435*	—	—	340	363*	—	—	—	—	—	—
54	337	267*	337	350*	—	—	—	—	337	360*	—	—	—	—	—	—
406x140x46	331	263*	—	—	—	—	—	—	—	—	—	—	—	—	—	—
39	327	261*	—	—	—	—	—	—	—	—	—	—	—	—	—	—
356x171x67	296	240*	296	315*	296	390	296	464	296	325*	296	442	—	—	—	—
57	292	237*	292	312*	292	386	—	—	292	321*	—	—	—	—	—	—
51	289	236*	289	310*	—	—	—	—	289	319*	—	—	—	—	—	—
45	282	231*	—	—	—	—	—	—	—	—	—	—	—	—	—	—
356x127x39	273	235*	—	—	—	—	—	—	—	—	—	—	—	—	—	—
33	—	—	—	—	—	—	—	—	—	—	—	—	—	—	—	—
305x165x54	244	206	244	272	244	337	—	—	244	280	—	—	—	—	—	—
46	241	204	241	270	—	—	—	—	241	278	—	—	—	—	—	—
40	234	193	—	—	—	—	—	—	—	—	—	—	—	—	—	—
305x127x48	244	206	229	242	—	—	—	—	227	246	—	—	—	—	—	—
42	232	189	213	216	—	—	—	—	210	219	—	—	—	—	—	—
37	221	170	—	—	—	—	—	—	—	—	—	—	—	—	—	—
305x102x33	215	157	—	—	—	—	—	—	—	—	—	—	—	—	—	—
28	—	—	—	—	—	—	—	—	—	—	—	—	—	—	—	—
25	—	—	—	—	—	—	—	—	—	—	—	—	—	—	—	—
254x146x43	193	169	180	190	—	—	—	—	177	193	—	—	—	—	—	—
37	183	144	—	—	—	—	—	—	—	—	—	—	—	—	—	—
31	—	—	—	—	—	—	—	—	—	—	—	—	—	—	—	—

398   Beam may be either grade S275 or grade S355
369   Beam must be grade S355 to satisfy neutral axis position requirements
264   Beam must be grade S275 to satisfy minimum reinforcement requirements (see Table B.1)
*   Reinforcement requires a guaranteed strain at maximum load of at least 10% for S355 beams, and possibly for S275 beams (check using Table B.1)
256   Connection capacity exceeds 0.8 M_p of composite beam in hogging (see ref. 16 Section 3.2.1 for significance of this)

The value of F_{r1} is based on the assumption that the NA is at least 200 mm below the bolt row. It should be reduced in accordance with ref. 16 Section 4.2 Step 1D when necessary.

Composite Connections

1 ROW M24 8.8 BOLTS 200 x 12 S275 END PLATE	COLUMN SIDE

S275											Column Serial Size	S355										
Panel Shear Cap.	Web Compn. Cap.	Tension Zone F_{r1}	Compression Zone									Compression Zone								Tension Zone F_{r1}	Web Compn. Cap.	Panel Shear Cap.
			Reinforcement option									Reinforcement option										
(kN)	(kN)	(kN)	A	B	C	D	E	F	G	H		A	B	C	D	E	F	G	H	(kN)	(kN)	(kN)
											356x368											
1000	1141	✓	✓	✓	✓	✓	S	✓	✓	S	S	x 202	✓	✓	✓	✓	✓	✓	✓	S	✓	1486	1302
849	935	✓	✓	✓	S	S	✓	S	S	S	S	x 177	✓	✓	✓	✓	✓	✓	S	S	✓	1217	1105
725	766	✓	✓	S	S	S	S	S	S	S	S	x 153	✓	✓	S	S	✓	S	S	S	✓	974	944
605	605	✓	S	S	S	S	S	S	S	S	S	x 129	✓	S	S	S	S	S	S	S	✓	788	787
											305x305											
1037	1432	✓	✓	✓	✓	✓	✓	✓	✓	✓	S	x198	✓	✓	✓	✓	✓	✓	✓	✓	✓	1865	1350
816	1051	✓	✓	✓	✓	S	✓	S	S	S	S	x158	✓	✓	✓	✓	✓	✓	S	S	✓	1368	1062
703	858	✓	✓	✓	S	S	S	S	S	S	S	x137	✓	✓	✓	S	✓	S	S	S	✓	1116	915
595	692	✓	✓	S	S	S	S	S	S	S	S	x118	✓	✓	S	S	✓	S	S	S	✓	909	774
503	553	✓	S	S	S	S	S	S	S	S	S	x97	✓	S	S	S	S	S	S	S	✓	713	649
											254x254											
882	1384	✓	✓	✓	✓	✓	✓	✓	✓	S	S	x167	✓	✓	✓	✓	✓	✓	✓	✓	✓	1802	1149
685	992	✓	✓	✓	S	S	✓	S	S	S	S	x132	✓	✓	✓	✓	✓	✓	S	S	✓	1292	892
551	744	✓	✓	S	S	S	S	S	S	S	S	x107	✓	✓	✓	S	✓	S	S	S	✓	696	717
434	557	✓	S	S	S	S	S	S	S	S	S	x89	✓	✓	S	S	✓	S	S	S	✓	725	566
360	436	297	S	S	S	S	S	S	S	S	S	x73	S	S	S	S	S	S	S	S	✓	563	465
											203x203											
459	701	✓	✓	S	S	S	S	S	S	S	S	x86	✓	✓	S	S	✓	S	S	S	✓	913	598
353	512	✓	S	S	S	S	S	S	S	S	S	x71	✓	S	S	S	S	S	S	S	✓	666	460
322	440	297	S	S	S	S	S	S	S	S	S	x60	S	S	S	S	S	S	S	S	✓	568	415

Tension Zone:
✓ Column satisfactory for bolt row tension values shown for the beam side.
195 Recalculate moment capacity based on reduced bolt row force (195 kN) using dimension 'A' to derive appropriate lever arm - or provide tension stiffener at the appropriate bolt row level.

Compression Zone:
✓ Column capacity exceeds $\Sigma F = F_{reinf} + F_{r1}$
S Provide compression stiffener.

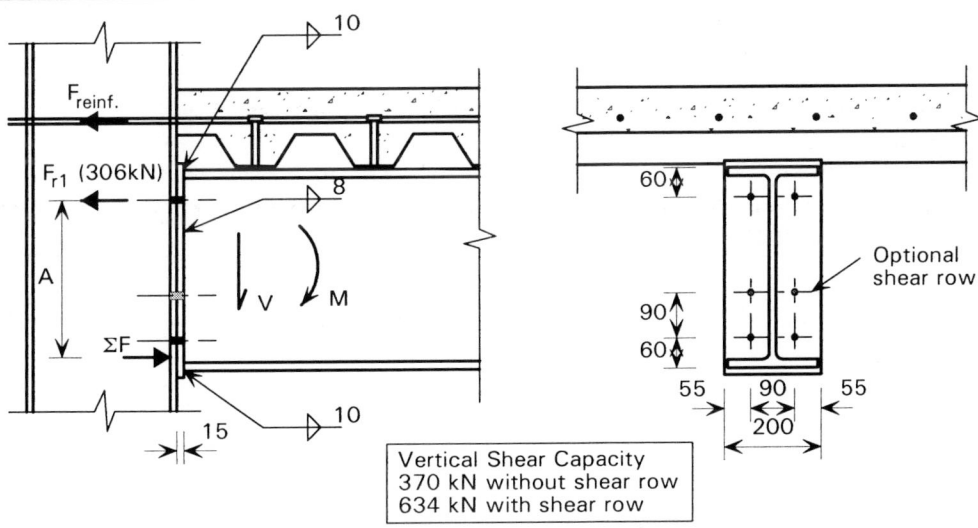

Vertical Shear Capacity
370 kN without shear row
634 kN with shear row

45

Composite Connections

	2 ROWS M24 8.8 BOLTS 200 x 15 S275 END PLATE							BEAM SIDE

BEAM Serial Size	Effective reinforcement (option, number and size of bars, A_{reinf}, F_{reinf})														
	A 4 No. φ16 804 mm² 351 kN		B 6 No. φ16 1210 mm² 529 kN		C 8 No. φ16 1610 mm² 704 kN		D 10 No. φ16 2010 mm² 878 kN		E 4 No. φ20 1260 mm² 551 kN		F 6 No. φ20 1890 mm² 826 kN		G 8 No. φ20 2510 mm² 1097 kN		H 10 No. φ20 3140 mm² 1372 kN	
	'A' mm	M_C kNm	'A' mm	M_C kNm	'A' mm	M_C kNm	'A' mm	M_C kNm	'A' mm	M_C kNm	'A' mm	M_C kNm	'A' mm	M_C kNm	'A' mm	M_C kNm
533x210x122	<u>384</u>	**445***	<u>384</u>	**552***	384	658*	384	763*	<u>384</u>	**566***	384	732*	384	895*	367	1029*
109	<u>380</u>	**442***	<u>380</u>	**548***	380	653*	380	758*	<u>380</u>	**561***	380	727*	366	*867**	**380**	**1054***
101	<u>378</u>	**440***	<u>378</u>	**546***	378	651*	*370*	*744**	<u>378</u>	**559***	373	716*	**378**	**886***	—	—
92	<u>375</u>	**437***	<u>375</u>	**543***	375	647*	*362*	*732**	<u>375</u>	**556***	365	706*	*370*	*873**	—	—
82	<u>372</u>	**434***	<u>367</u>	**534***	372	643*	*366*	*738**	<u>366</u>	**546***	372	715*	—	—	—	—
457x191x98	<u>308</u>	**377***	308	471*	308	563*	*302*	*647**	308	483*	308	628*	289	*734**	**299**	900
89	<u>305</u>	**375***	305	468*	305	560*	*291*	*631**	305	479*	294	*610**	**305**	**766***	—	—
82	<u>302</u>	**373***	302	465*	294	*547**	*282*	*610**	302	477*	286	*594**	—	—	—	—
74	<u>300</u>	**371***	295	458*	300	554*	*293*	*635**	294	468*	**300**	**617***	—	—	—	—
67	<u>297</u>	**368***	283	*444**	*291*	*543**	—	—	282	450*	—	—	—	—	—	—
457x152x82	<u>306</u>	**376***	306	470*	*293*	*545**	*279*	*600**	306	481*	283	*585**	*292*	*745**	—	—
74	<u>304</u>	**374***	294	456*	*279*	*516**	*292*	*634**	292	466*	267	*549**	—	—	—	—
67	<u>301</u>	**371***	284	*443**	*292*	*544**	—	—	282	450*	—	—	—	—	—	—
60	<u>287</u>	**359***	*291*	*454**	—	—	—	—	290	464*	—	—	—	—	—	—
52	<u>267</u>	**327***	—	—	—	—	—	—	—	—	—	—	—	—	—	—
406x178x74	255	331*	255	415*	*242*	*469**	*228*	*510*	255	425*	232	*503**	—	—	—	—
67	252	328*	243	*394**	*228*	*441**	—	—	241	400*	**244**	**533***	—	—	—	—
60	245	320*	231	*369**	—	—	—	—	228	*374**	—	—	—	—	—	—
54	232	*296**	*212*	*322**	—	—	—	—	*209*	*323**	—	—	—	—	—	—
406x140x46	213	*251**	—	—	—	—	—	—	—	—	—	—	—	—	—	—
39	—	—	—	—	—	—	—	—	—	—	—	—	—	—	—	—

398   Beam may be either grade S275 or grade S355
369   Beam must be grade S355 to satisfy neutral axis position requirements
<u>264</u>   Beam must be grade S275 to satisfy minimum reinforcement requirements (see Table B.1)
*   Reinforcement requires a guaranteed strain at maximum load of at least 10% for S355 beams, and possibly for S275 beams (check using Table B.1)
256   Connection capacity exceeds 0.8 M_p of composite beam in hogging (see ref. 16 Section 3.2.1 for significance of this)
The value of F_{r1} is based on the assumption that the NA is at least 200 mm below the bolt row. It should be reduced in accordance with ref. 16 Section 4.2 Step 1D when necessary.

Composite Connections

2 ROWS M24 8.8 BOLTS
200 x 15 S275 END PLATE — COLUMN SIDE

S275												Column Serial Size	S355											
Panel Shear Cap.	Web Compn. Cap.	Tension Zone		Compression Zone									Compression Zone								Tension Zone		Web Compn. Cap.	Panel Shear Cap.
		F_{r1}	F_{r2}	Reinforcement option									Reinforcement option								F_{r1}	F_{r2}		
(kN)	(kN)	(kN)		A	B	C	D	E	F	G	H		A	B	C	D	E	F	G	H	(kN)		(kN)	(kN)
												356x368												
1000	1141	✓	✓	✓	✓	S	S	✓	S	S	S	x 202	✓	✓	✓	✓	✓	✓	S	S	✓	✓	1486	1302
849	935	✓	✓	✓	S	S	S	S	S	S	S	x 177	✓	✓	S	S	✓	S	S	S	✓	✓	1217	1105
725	766	✓	✓	S	S	S	S	S	S	S	S	x 153	✓	S	S	S	S	S	S	S	✓	✓	974	944
605	605	✓	✓	S	S	S	S	S	S	S	S	x 129	S	S	S	S	S	S	S	S	✓	✓	788	787
												305x305												
1037	1432	✓	✓	✓	✓	✓	✓	✓	✓	S	S	x198	✓	✓	✓	✓	✓	✓	✓	S	✓	✓	1865	1350
816	1051	✓	✓	✓	S	S	S	S	S	S	S	x158	✓	✓	✓	S	✓	S	S	S	✓	✓	1368	1062
703	858	✓	✓	S	S	S	S	S	S	S	S	x137	✓	✓	S	S	✓	S	S	S	✓	✓	1116	915
595	692	✓	✓	S	S	S	S	S	S	S	S	x118	✓	S	S	S	S	S	S	S	✓	✓	909	774
503	553	✓	✓	S	S	S	S	S	S	S	S	x97	S	S	S	S	S	S	S	S	✓	✓	713	649
												254x254												
882	1384	✓	✓	✓	✓	✓	S	✓	✓	S	S	x167	✓	✓	✓	✓	✓	✓	✓	S	✓	✓	1802	1149
685	992	✓	✓	✓	S	S	S	S	S	S	S	x132	✓	✓	✓	S	✓	S	S	S	✓	✓	1292	892
551	744	✓	✓	S	S	S	S	S	S	S	S	x107	✓	S	S	S	S	S	S	S	✓	✓	969	717
434	557	✓	✓	S	S	S	S	S	S	S	S	x89	S	S	S	S	S	S	S	S	✓	✓	725	566
360	436	297	✓	S	S	S	S	S	S	S	S	x73	S	S	S	S	S	S	S	S	✓	✓	563	465
												203x203												
459	701	✓	✓	S	S	S	S	S	S	S	S	x86	✓	S	S	S	S	S	S	S	✓	✓	913	598
353	512	✓	✓	S	S	S	S	S	S	S	S	x71	S	S	S	S	S	S	S	S	✓	✓	666	460
322	440	297	204	S	S	S	S	S	S	S	S	x60	S	S	S	S	S	S	S	S	✓	✓	568	415

Tension Zone:
✓ Column satisfactory for bolt row tension values shown for the beam side.
195 Recalculate moment capacity based on reduced bolt row force (195 kN) using dimension 'A' to derive appropriate lever arm - or provide tension stiffener at the appropriate bolt row level.

Compression Zone:
✓ Column capacity exceeds $\Sigma F = F_{reinf} + F_{r1}$
S Provide compression stiffener.

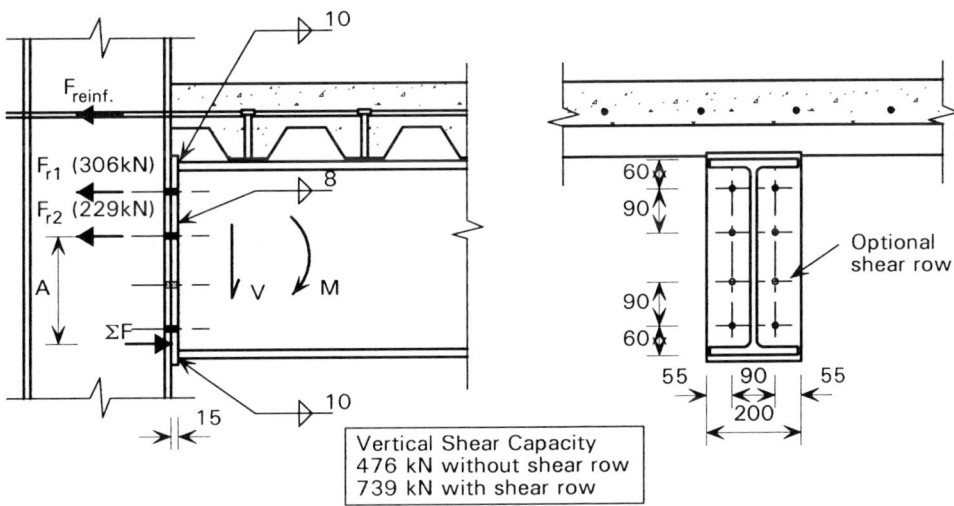

Vertical Shear Capacity
476 kN without shear row
739 kN with shear row

Composite Connections

2 ROWS M24 8.8 BOLTS 250 x 15 S275 END PLATE	BEAM SIDE

BEAM Serial Size	Effective reinforcement (option, number and size of bars, A_{reinf}, F_{reinf})															
	A 4 No. φ16 804 mm² 351 kN		B 6 No. φ16 1210 mm² 529 kN		C 8 No. φ16 1610 mm² 704 kN		D 10 No. φ16 2010 mm² 878 kN		E 4 No. φ20 1260 mm² 551 kN		F 6 No. φ20 1890 mm² 826 kN		G 8 No. φ20 2510 mm² 1097 kN		H 10 No. φ20 3140 mm² 1372 kN	
	'A' mm	M_C kNm	'A' mm	M_C kNm	'A' mm	M_C kNm	'A' mm	M_C kNm	'A' mm	M_C kNm	'A' mm	M_C kNm	'A' mm	M_C kNm	'A' mm	M_C kNm
533x210x122	384	459*	384	566*	384	671*	384	777*	384	579*	384	745*	384	909*	365	1038*
109	380	455*	380	562*	380	666*	380	771*	380	575*	380	740*	364	876*	374	1055*
101	378	453*	378	559*	378	664*	368	754*	378	572*	371	727*	378	899*	–	–
92	375	451*	375	556*	370	653*	359	741*	375	569*	363	716*	369	884*	–	–
82	372	447*	365	545*	353	632*	364	748*	364	557*	343	721*	–	–	–	–
457x191x98	308	388*	308	482*	308	574*	300	655*	308	493*	308	639*	287	736*	298	908
89	305	385*	305	478*	299	563*	288	633*	305	490*	292	617*	299	766*	–	–
82	302	383*	302	476*	292	554*	279	610*	302	487*	283	596*	–	–	–	–
74	300	381*	293	466*	280	529*	292	643*	292	476*	294	620*	–	–	–	–
67	297	379*	280	446*	289	551*	–	–	278	451*	–	–	–	–	–	–
457x152x82	306	387*	301	475*	290	550*	276	599*	300	485*	280	585*	–	–	–	–
74	304	384*	291	464*	276	516*	290	641*	289	473*	263	546*	–	–	–	–
67	295	376*	281	445*	290	552*	–	–	279	451*	–	–	–	–	–	–
60	283	365*	289	462*	–	–	–	–	288	472*	–	–	–	–	–	–
52	262	326*	–	–	–	–	–	–	–	–	–	–	–	–	–	–

398   Beam may be either grade S275 or grade S355
369   Beam must be grade S355 to satisfy neutral axis position requirements
264   Beam must be grade S275 to satisfy minimum reinforcement requirements (see Table B.1)
*   Reinforcement requires a guaranteed strain at maximum load of at least 10% for S355 beams, and possibly for S275 beams (check using Table B.1)
256   Connection capacity exceeds 0.8 M_p of composite beam in hogging (see ref. 16 Section 3.2.1 for significance of this)
The value of F_{r1} is based on the assumption that the NA is at least 200 mm below the bolt row. It should be reduced in accordance with ref. 18 Section 4.2 Step 1D when necessary.

Composite Connections

2 ROWS M24 8.8 BOLTS
250 x 15 S275 END PLATE
COLUMN SIDE

Panel Shear Cap. (kN)	Web Compn. Cap. (kN)	Tension Zone S275 F_{r1}	F_{r2}	Compression Zone — Reinforcement option A	B	C	D	E	F	G	H	Column Serial Size	Compression Zone — Reinforcement option A	B	C	D	E	F	G	H	Tension Zone S355 F_{r1}	F_{r2}	Web Compn. Cap. (kN)	Panel Shear Cap. (kN)
		(kN)																			(kN)			
												356x368												
1000	1141	✓	✓	✓	✓	S	S	✓	S	S	S	x 202	✓	✓	✓	✓	✓	✓	S	S	✓	✓	1486	1302
849	935	✓	✓	✓	S	S	S	S	S	S	S	x 177	✓	✓	S	S	✓	S	S	S	✓	✓	1217	1105
725	766	✓	✓	S	S	S	S	S	S	S	S	x 153	✓	S	S	S	S	S	S	S	✓	✓	974	944
605	605	✓	✓	S	S	S	S	S	S	S	S	x 129	S	S	S	S	S	S	S	S	✓	✓	788	787
												305x305												
1037	1432	✓	✓	✓	✓	✓	S	✓	✓	S	S	x198	✓	✓	✓	✓	✓	✓	✓	S	✓	✓	1865	1350
816	1051	✓	✓	✓	S	S	S	S	S	S	S	x158	✓	✓	✓	S	✓	S	S	S	✓	✓	1368	1062
703	858	✓	✓	S	S	S	S	S	S	S	S	x137	✓	✓	S	S	S	S	S	S	✓	✓	1116	915
595	692	✓	✓	S	S	S	S	S	S	S	S	x118	S	S	S	S	S	S	S	S	✓	✓	909	774
503	553	✓	✓	S	S	S	S	S	S	S	S	x97	S	S	S	S	S	S	S	S	✓	✓	713	649
												254x254												
882	1384	✓	✓	✓	✓	✓	S	✓	S	S	S	x167	✓	✓	✓	✓	✓	✓	S	S	✓	✓	1802	1149
685	992	✓	✓	✓	S	S	S	S	S	S	S	x132	✓	✓	✓	S	✓	S	S	S	✓	✓	1292	892
551	744	✓	✓	S	S	S	S	S	S	S	S	x107	✓	S	S	S	S	S	S	S	✓	✓	969	717
434	557	✓	✓	S	S	S	S	S	S	S	S	x89	S	S	S	S	S	S	S	S	✓	✓	725	566
360	436	297	✓	S	S	S	S	S	S	S	S	x73	S	S	S	S	S	S	S	S	✓	✓	563	465
												203x203												
459	701	✓	✓	S	S	S	S	S	S	S	S	x86	S	S	S	S	S	S	S	S	✓	✓	913	598
353	512	✓	✓	S	S	S	S	S	S	S	S	x71	S	S	S	S	S	S	S	S	✓	✓	666	460
322	440	297	204	S	S	S	S	S	S	S	S	x60	S	S	S	S	S	S	S	S	✓	✓	568	415

Tension Zone:
✓ Column satisfactory for bolt row tension values shown for the beam side.
195 Recalculate moment capacity based on reduced bolt row force (195 kN) using dimension 'A' to derive appropriate lever arm - or provide tension stiffener at the appropriate bolt row level.

Compression Zone:
✓ Column capacity exceeds $\Sigma F = F_{reinf} + F_{r1}$
S Provide compression stiffener.

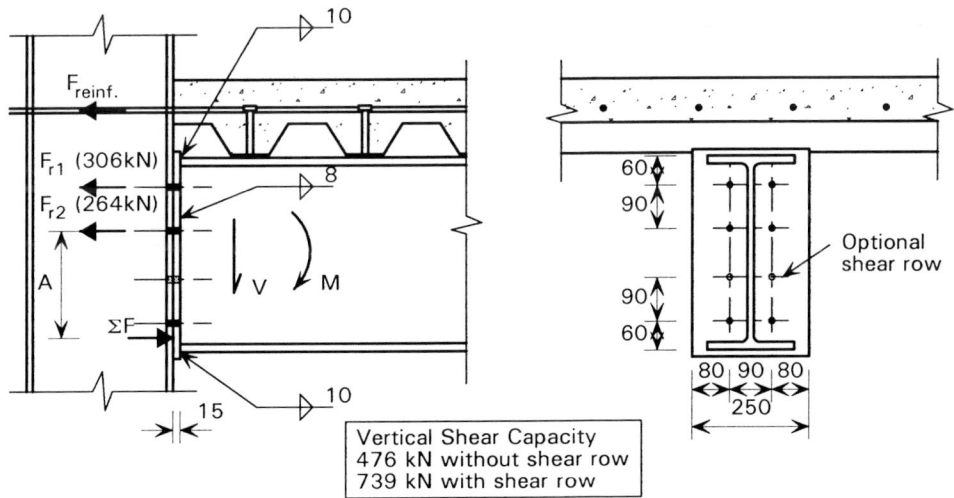

Vertical Shear Capacity
476 kN without shear row
739 kN with shear row

Composite Connections

DIMENSIONS FOR DETAILING

Beam serial size	Dimension a_1 mm	Dimension a_2 mm	End plate overall depth D_F mm
533x210x122	425	245	
109	420	240	
101	415	235	600
92	415	235	
82	410	230	
457x191x98	350	170	
89	345	165	
82	340	160	520
74	340	160	
67	335	155	
457x152x82	345	165	
74	340	160	
67	340	160	520
60	335	155	
52	330	150	
406x178x74	295	115	
67	290	110	470
60	285	105	
54	285	105	
406x140x46	280	100	450
39	275	95	
356x171x67	245	—	
57	240	—	420
51	235	—	
45	230	—	
356x127x39	235	—	410
33	230	—	
305x165x54	190	—	
46	185	—	360
40	185	—	
305x127x48	190	—	
42	185	—	360
37	185	—	
305x102x33	195	—	
28	190	—	370
25	185	—	
254x146x43	140	—	
37	135	—	310
31	135	—	
254x102x28	140	—	
25	135	—	310
22	135	—	

See capacity table diagram for plate thickness and other dimensions appropriate to the moment capacities. All plates to be S275.

APPENDIX D: Worked Example

D.1 Introduction

The design example given in this Appendix is for a frame that is braced out of plane in order to prevent sway about the minor axis of the columns.

Calculations are given to demonstrate the following aspects of the design rules:

- compliance with the scope
- framing and loads
- wind analysis
- notional horizontal forces and analysis
- beam design
- column loads
- internal column design
- external column design
- connections design
- serviceability limit state.

D.1.1 Compliance

The frame in this example forms part of a steel structure that conforms to the frame layout specified in Section 1.2. In particular:

- The floor layout comprises primary and secondary beams as shown in Figure 1.3
- The frame is braced against sway about the minor axis
- The bay width is constant over the frame height

D.1.2 Frame dimensions

The frame dimensions conform to the limitations given in Section 1.2.1:

No. of storeys	=	4			OK
No. of bays	=	5	Therefore consider only 4 to resist wind loads		
Bay width	=	9 m			OK
Bottom storey height	=	4.5 m			OK
Storey height (above bottom storey)	=	3.5 m			OK
Bay width / storey height (bottom)	=	9 / 4.5	=	2	OK
Bay width / storey height (above bottom storey)	=	9 / 3.5	=	2.57	OK
Greatest bay width / smallest bay width	=		9 / 9 =	1	OK

D.1.3 Loading

The following unfactored loading conforms to the range given in Section 1.3.3, Table 2:

Dead load on floors	= 5 kN/m^2	OK
Imposed load on floors	= 7.5 kN/m^2	OK
Dead load on roof	= 3.75 kN/m^2	OK
Imposed load on roof	= 1.5 kN/m^2	OK

Wind loads see Figure D.2

D.1.4 Design

The members were designed using rules given in BS 5950-1:1990[2] and BS 5950-3: 1990[3], with additional requirements as specified in this publication.

Member capacities were either obtained directly from tables given in *Steelwork design guide to BS 5950: Part 1:1990 - Volume 1*[18] or calculated using the method given in BS 5950-3.

The Steel Construction Institute 		Job No:	**P264**	Page	**1** of **25**	Rev **A**
		Job Title	**Wind-moment Composite Frames**			
		Subject	**Framing and Loads**			
Silwood Park, Ascot, Berks SL5 7QN Telephone: (01344) 623345 Fax: (01344) 622944		Client	**SCI**	Made by	**JSH**	Date **Nov. 99**
CALCULATION SHEET				Checked by	**AW**	Date **Dec. 99**

D.2 FRAMING AND LOADS

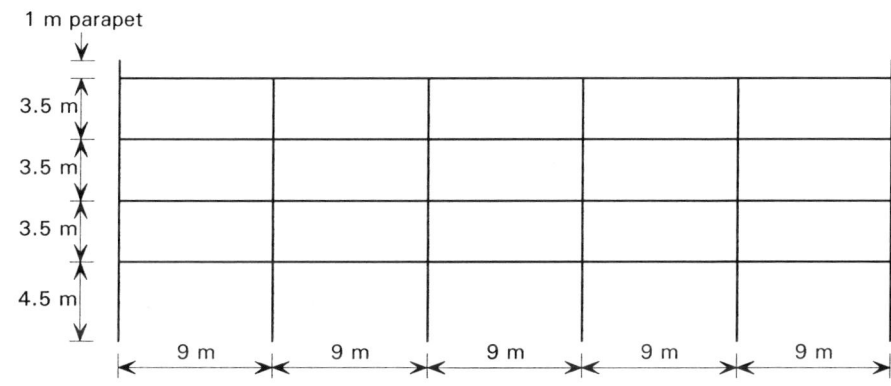

Frame spacing is at 6m centres

Figure D.1 Frame elevation

In accordance with Section 1.3.1, the number of active bays in the wind-moment frame must not exceed four. Therefore, consider only four of the five bays to be active when carrying out the wind analysis using the portal method, see below.

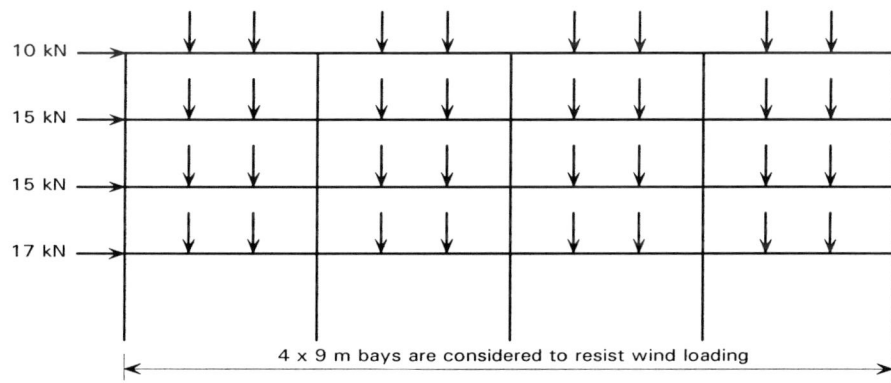

Beams are loaded by transverse beams at ⅓ span points

Figure D.2 Frame loading arrangement

Roof
Dead load $3.75 kN/m^2$ equivalent to two point loads of $67.5 kN/m^2$
Imposed load $1.50 kN/m^2$ equivalent to two point loads of $27 kN/m^2$

Floor
Dead load $5.00 kN/m^2$ equivalent to two point loads of $90 kN/m^2$
Imposed load $7.50 kN/m^2$ equivalent to two point loads of $135 kN/m^2$

The Steel Construction Institute 	Job No: **P264**		Page **2** of **25**	Rev **A**
	Job Title	**Wind-moment Composite Frames**		
	Subject	**Wind Analysis**		
Silwood Park, Ascot, Berks SL5 7QN Telephone: (01344) 623345 Fax: (01344) 62944				
	Client **SCI**	Made by **JSH**	Date	**Nov. 99**
CALCULATION SHEET		Checked by **AW**	Date	**Dec. 99**

D.3 WIND ANALYSIS

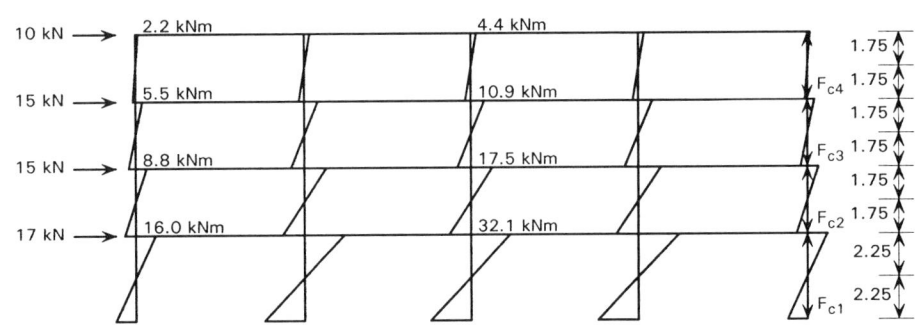

Figure D.3 Frame analysis under wind loads (columns)

Table D.1 Shear forces and bending moments in columns due to wind load

Storey	Total wind shear (kN)	Shear force in column (kN)		Bending moment in column (kNm)	
		External	Internal	External	Internal
4	10	1.25	2.5	$1.25 \times 1.75 = 2.2$	$2.5 \times 1.75 = 4.4$
3	25	3.13	6.25	$3.13 \times 1.75 = 5.5$	$6.25 \times 1.75 = 10.9$
2	40	5	10	$5.0 \times 1.75 = 8.8$	$10.0 \times 1.75 = 17.5$
1	57	7.13	14.25	$7.13 \times 2.25 = 16$	$14.25 \times 2.25 = 32.1$

Table D.2 Moments and axial loads due to wind loads

Storey	Moments about point of contraflexure at mid-height	F_c (kN)
4	$36 F_{c4} = 10 \times 1.75$	0.5
3	$36 F_{c3} = 10 \times 5.25 + 15 \times 1.75$	2.2
2	$36 F_{c2} = 10 \times 8.75 + 15 \times 5.25 + 15 \times 1.75$	5.3
1	$36 F_{c1} = 10 \times 12.75 + 15 \times 9.25 + 15 \times 5.75 + 17 \times 2.25$	10.9

N.B. Values are UNFACTORED

Axial forces in the beams due to wind loads are small and may be neglected.

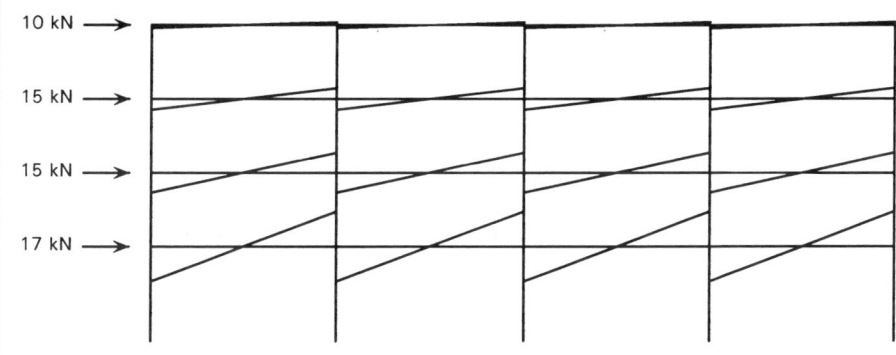

Figure D.4 Frame analysis under wind loads (beams)

Table D.3 Bending moments in beams due to wind load

Floor level	Bending moment in external Column (kNm)		Bending moment in beam (kNm)
	Upper column	Lower column	
4	0	2.2	0.0 + 2.2 = 2.2
3	2.2	5.5	2.2 + 5.5 = 7.7
2	5.5	8.8	5.5 + 8.8 = 14.3
1	8.8	16	8.8 + 16.0 = 24.8

N.B. Values are UNFACTORED

D.4 NOTIONAL HORIZONTAL FORCES

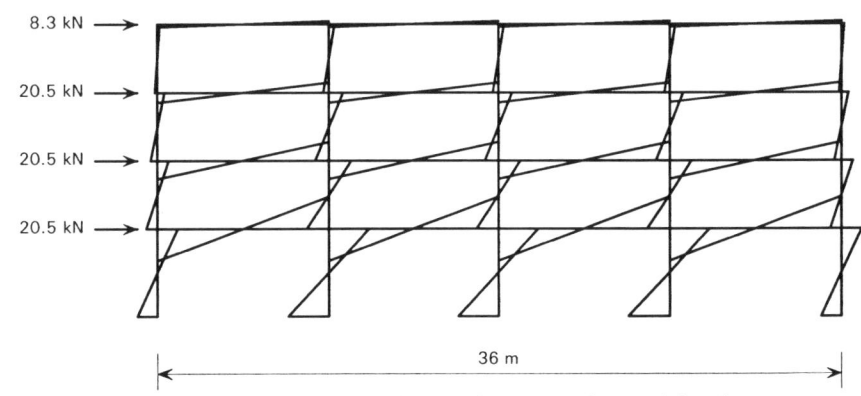

Figure D.5 Frame analysis under notional horizontal loads

Notional horizontal force = 0.005 (1.4 dead + 1.6 imposed)

Roof H = 0.005 (1.4 × 3.75 + 1.6 × 1.5) × 36 × 6 = 8.3 kN

Floor H = 0.005 (1.4 × 5.0 + 1.6 × 7.5) × 36 × 6 = 20.5 kN

Table D.4 Shear force and bending moment in the columns due to notional horizontal forces

Storey	Total shear (kN)	Shear force in column (kN)		Bending moment in column (kNm)	
		External	Internal	External	Internal
4	8.3	1	2.1	1.0 × 1.75 = 1.8	2.1 × 1.75 = 3.6
3	28.8	3.6	7.2	3.6 × 1.75 = 6.3	7.2 × 1.75 = 12.6
2	49.3	6.2	12.3	6.2 × 1.75 = 10.9	12.3 × 1.75 = 21.5
1	69.8	8.7	17.5	8.7 × 2.25 = 19.6	17.5 × 2.25 = 39.4

Table D.5 Bending moments in beams due to notional horizontal forces

Floor level	Bending moment in external column (kNm)		Bending moment in beam (kNm)
	Upper column	Lower column	
Roof	0	1.8	0.0 + 1.8 = 1.8
3	1.8	6.3	1.8 + 6.3 = 8.1
2	6.3	10.9	6.3 + 10.9 = 17.2
1	10.9	19.6	10.9 + 19.6 = 30.5

N.B Values are UNFACTORED

The Steel Construction Institute 		Job No:	BCB478		Page	5 of 25	Rev A
		Job Title	Wind-moment Composite Frames				
Silwood Park, Ascot, Berks SL5 7QN Telephone: (01344) 623345 Fax: (01344) 622944		Subject	Roof Beam Design				
		Client	SCI	Made by	JSH	Date	Nov. 99
CALCULATION SHEET				Checked by	AW	Date	Dec. 99

D.5 BEAM DESIGN

D.5.1 Roof beam design

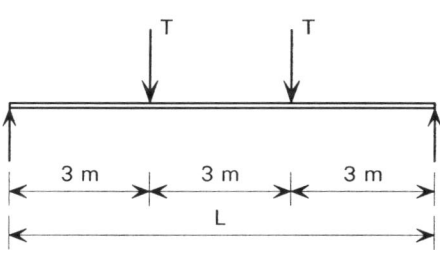

Figure D.6 Roof beam

<u>Ultimate Limit State</u>

<u>Dead and Imposed Loading</u>

(T=point loads from transverse beams)

Design load for ULS: $T = 1.4 \times 67.5 + 1.6 \times 27 = 137.7$ kN   D.2, Sheet 1

Moment:
Take end restraint moment due to gravity load equal to 10% of the maximum sagging moment of a simply-supported beam   Section 4.1.2

Sagging moment:
$$M = \frac{0.9TL}{3} = \frac{0.9 \times 137.7 \times 9}{3} = 372 \text{ kNm}$$

Hogging moment:
$$M = \frac{0.1TL}{3} = \frac{0.1 \times 137.7 \times 9}{3} = 41.3 \text{ kNm}$$

Shear:
$$F_v = (1.4 \times 3.75 + 1.6 \times 1.50) \times 6 \times 9/2 = 207 \text{ kN}$$ Sheet 1

<u>Try 356 × 171 UB 57 Grade S355</u>

Slab depth = 140 mm

Metal decking = 1 mm gauge, 50 mm deep, trapezoidal profile

Shear studs = 19 mm φ, 95 mm high @ 150 mm c/c

Longitudinal rebar = T16@ 150 mm

The Steel Construction Institute  Silwood Park, Ascot, Berks SL5 7QN Telephone: (01344) 623345 Fax: (01344) 622944 **CALCULATION SHEET**	Job No: **BCB478**		Page **6** of **25**	Rev **A**
	Job Title	**Wind-moment Composite Frames**		
	Subject	**Roof Beam Design**		
	Client **SCI**	Made by **JSH**	Date	**Nov. 99**
		Checked by **AW**	Date	**Dec. 99**

Transverse rebar = T12 @ 200 mm	
Concrete = C30 lightweight concrete	
The composite beam is designed using BS 5950-3: Section 3.1: 1990, and is classified as a Class 1 plastic section.	Ref. 3 Ref. 18
Section sagging bending capacity, M_p = 680 kNm	Ref. 3
Reduced composite beam sagging capacity:	
$\quad$ 0.9 M_p = 612 kNm > 372 kNm $\qquad$ OK	
Beam hogging moment capacity, based on steel section alone:	
$\quad M_{hog}$ = 359 kNm > 41.3 kNm $\qquad$ OK	Ref. 18
Shear check: P_v = 0.6 $p_y A_v$ = 618 kN > 207 kN $\qquad$ OK	Ref. 18
<u>Dead load plus wind loading</u>	
Maximum roof beam <u>end moments</u> due to wind = 2.2 kNm	Sheet 3
Maximum hogging moment = $1.4 \left(\dfrac{0.1 \times 67.5 \times 9}{3} + 2.2 \right)$ = 31.4 kNm	Sheet 1
Hence, wind load does not cause sagging moments at beam ends	
and $\quad M_{hog}$ > 31.4 kNm $\qquad$ OK	
<u>Dead load plus imposed load plus wind loading</u>	
Maximum roof beam <u>end moments</u> due to wind = 2.2 kNm	
Maximum hogging moment = $1.2 \left(\dfrac{0.1 \times (67.5 + 27) \times 9}{3} + 2.2 \right)$	
$\qquad\qquad\qquad\qquad$ = 36.7 kNm	
Hence, wind load does not cause sagging moments at beam end	
and $\quad M_{hog}$ > 36.7 kNm $\qquad$ OK	

The Steel Construction Institute 	Job No: **BCB478**		Page **7** of **25**	Rev **A**
	Job Title	**Wind-moment Composite Frames**		
Silwood Park, Ascot, Berks SL5 7QN Telephone: (01344) 623345 Fax: (01344) 622944	Subject	**Roof Beam Design**		
	Client **SCI**	Made by **JSH**	Date	**Nov. 99**
CALCULATION SHEET		Checked by **AW**	Date	**Dec. 99**

Serviceability limit state

Design load for SLS: $T = 27$ kN

Sheet 1

Deflection of simply supported beam:

$$\delta_{mid} = \frac{2Ta(3L^2 - 4a^2)}{48\,EI} \quad \text{where} \quad a = \frac{L}{3}$$

for unpropped construction with partial shear connection

$$\delta = \delta_c + 0.3(1 - N_a/N_p)(\delta_s - \delta_c)$$

Ref. 3
Cl. 6.1.4

where: δ_c = deflection of uncracked composite section
δ_s = deflection of steel section acting alone

$E = 205$ kN/mm^2
$I_{steel} = 1604 \times 10^5$ mm^4
$I_{compo} = 5238 \times 10^5$ mm^4

Ref. 2
Ref. 18
Ref. 3

$$\delta_c = \frac{2 \times 27 \times 3 \times 10^3 \,[3\,(9 \times 10^3)^2 - 4\,(3 \times 10^3)^2]}{48 \times 205 \times 5238 \times 10^5}$$

$= 6.5$ mm

$$\delta_s = \frac{2 \times 27 \times 3 \times 10^3 \,[3\,(9 \times 10^3)^2 - 4\,(3 \times 10^3)^2]}{48 \times 205 \times 1604 \times 10^5}$$

$= 21.25$ mm

Deflection at mid-span:

$$\delta = \delta_c + \left(1 - \frac{29}{36}\right)(21.25 - 6.5) = 7.36\,\text{mm}$$

$= \dfrac{span}{1223} < \dfrac{span}{360}$ OK

Use 356 ×171 ×57 UB composite beam

	Job No: **BCB478**	Page **8** of **25**	Rev **A**
The Steel Construction Institute  Silwood Park, Ascot, Berks SL5 7QN Telephone: (01344) 623345 Fax: (01344) 622944	Job Title **Wind-moment Composite Frames**		
	Subject **Floor Beam Design**		
CALCULATION SHEET	Client **SCI**	Made by **JSH**	Date **Nov. 99**
		Checked by **AW**	Date **Dec. 99**

D.5.2 Floor beam design

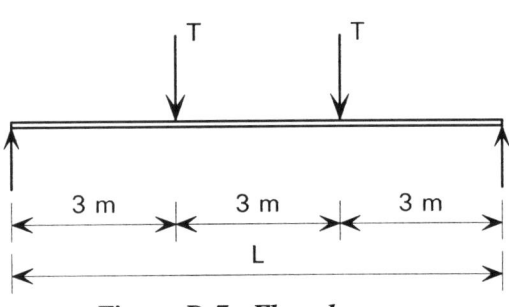

Figure D.7 Floor beam

<u>Ultimate Limit State</u>

<u>Dead and Imposed Loading</u>

(T = point loads from transverse beam)

Design load for ULS: $T = 1.4 \times 90 + 1.6 \times 135 = 342$ kN		Sheet 1

Moment:
Take end restraint moments due to gravity load equal to 10% of the maximum sagging moment of a simply supported beam   Section 4.1.2

Sagging moment:

$$M = \frac{0.9\ TL}{3} = \frac{0.9 \times 342 \times 9}{3} = 923 \text{ kNm}$$

Hogging moment:

$$M = \frac{0.1\ TL}{3} = \frac{0.1 \times 342 \times 9}{3} = 103 \text{ kNm}$$

Shear:

$$F_v = (1.4 \times 5.00 + 1.6 \times 7.50) \times 6 \times 9/2 = 513 \text{ kN}$$ Sheet 1

<u>Try 457 × 191 UB 98 Grade S355</u>

Slab depth   = 140 mm

Metal decking = 1 mm gauge, 50 mm deep, trapezoidal profile

The Steel Construction Institute	Job No: **BCB478**		Page **9** of **25**	Rev **A**
Silwood Park, Ascot, Berks SL5 7QN Telephone: (01344) 623345 Fax: (01344) 622944	Job Title **Wind-moment Composite Frames**			
	Subject **Floor Beam Design**			
CALCULATION SHEET	Client **SCI**	Made by **JSH**	Date **Nov. 99**	
		Checked by **AW**	Date **Dec. 99**	

Shear studs = 19 mm φ, 95 mm high @ 150 mm c/c

Longitudinal rebar = T16 @ 150 mm

Transverse rebar = T12 @ 200 mm

Concrete = C30 lightweight concrete

The composite beam is designed using BS 5950-3: Section 3.1: 1990, and is classified as a Class 1 plastic section. *Ref. 3* *Ref. 18*

Section sagging moment capacity, M_p = 1211 kNm *Ref. 3*

Reduced composite beam sagging capacity:

$$0.9\, M_p = 1090 \text{ kNm} > 923 \text{ kNm} \quad \text{OK}$$

Beam hogging moment capacity, based on steel section alone:

$$M_{hog} = 770 \text{ kNm} > 103 \text{ kNm} \quad \text{OK} \quad \textit{Ref. 18}$$

Shear check: $P_v = 0.6\, p_y A_v$ = 1100 kN > 513 kN   OK   *Ref. 18*

<u>Dead load plus wind loading</u>

Maximum floor beam <u>end moments</u> due to wind = 24.8 kNm   *Sheet 3*

$$\text{Maximum hogging moment} = 1.4 \left(\frac{0.1 \times 90 \times 9}{3} + 24.8 \right) = 72.5 \text{ kNm} \quad \textit{Sheet 1}$$

Hence, wind load does not cause sagging moments at beam ends

and $M_{hog} > 72.5$ kNm   OK

<u>Dead load plus imposed load plus wind loading</u>

Maximum floor beam <u>end moments</u> due to wind = 24.8 kNm   *Sheet 3*

$$\text{Maximum hogging moment} = 1.2 \left(\frac{0.1 \times (90 + 135) \times 9}{3} + 24.8 \right)$$ *Sheet 1*

$$= 111 \text{ kNm}$$

Hence, wind load does not cause sagging moments at beam end

The Steel Construction Institute  Silwood Park, Ascot, Berks SL5 7QN Telephone: (01344) 623345 Fax: (01344) 622944 **CALCULATION SHEET**	Job No: **BCB478**		Page **10** of **25**	Rev **A**
	Job Title	**Wind-moment Composite Frames**		
	Subject	**Floor Beam Design**		
	Client **SCI**	Made by **JSH**	Date **Nov. 99**	
		Checked by **AW**	Date **Dec. 99**	

and $M_{hog} > 111\ kNm$ OK

<u>Serviceability limit state</u>

Design load for SLS: $T = 135\ kN$ *Sheet 1*

Deflection checked for unfactored imposed load (similar to Sheet 7)   OK

<u>Use 457 × 191 × 98 UB composite beam</u>

D.6 COLUMN LOADS

Data for calculation of column moments are given in Table D.6.

Table D.6   Data for calculation of column moments

Storey	Beam reactions (kN)		10% restraint Moment (kNm)		Moments due to horizontal loads (kNm)			
					Notional loads		Wind loads	
	Dead (kN)	Imposed (kN)	Dead (kNm)	Imposed (kNm)	External (kNm)	Internal (kNm)	External (kNm)	Internal (kNm)
3	90	135	27	40.5	6.3	12.6	5.5	10.9
1	90	135	27	40.5	19.6	39.4	16	32.1

Sheet 1

Table D.4

Table D.1

N.B.   All values are UNFACTORED, except for moments due to notional horizontal loads

The values for the 10% restraint moment are calculated from the unfactored floor loads

Dead   = $0.1 \times 90 \times 9 / 3$ =   27.0 kNm

Imposed   = $0.1 \times 135 \times 9 / 3$ =   40.5 kNm

The Steel Construction Institute 			Job No:	BCB478	Page 11 of 25		Rev A
			Job Title	Wind-moment Composite Frames			
			Subject	Internal Columns			
Silwood Park, Ascot, Berks SL5 7QN Telephone: (01344) 623345 Fax: (01344) 622944			Client	SCI	Made by	JSH	Date Nov. 99
CALCULATION SHEET					Checked by	AW	Date Dec. 99

D.7 INTERNAL COLUMNS

The columns will be spliced above the second storey floor beams, where a change in section size may take place. Therefore, design calculations will be required for storeys 3 and 1.

Table D.7 Loading on internal columns

Storey	Loading (kN) Dead/Imposed — Dead/Imposed	S_w of col. (kN)	Total load (kN) Dead	Total load (kN) Imposed	Reduction in imposed load (kN)	Reduced imposed load (kN)
4	101.3/40.5 — 101.3/40.5	3	206	81	0% / 0	81
3	135.0/202.5 — 135.0/202.5	3	479	486	10% / 49	437
2	135.0/202.5 — 135.0/202.5	5	754	891	20% / 178	713
1	135.0/202.5 — 135.0/202.5	6	1030	1296	30% / 389	907

N.B. Values are UNFACTORED

The reduction in imposed load for the number of storeys carried is given by BS 6399-1: Table 2.

D.7.1 Storey 3

<u>Dead load plus imposed load plus notional horizontal forces</u>

Design load at ULS: $F_c = 1.4 \times 479 + 1.6 \times 437 = 1370$ kN

Design moment at ULS: M_x = 12.6 kNm   *Table D.6*

Moments due to eccentric reactions and the 10% restraint moment balance and produce no net moment about the major axis.

Storey height = 3.5 m

L_{Ey} = 1.0L = 3.5 m, L_{Ex} = 1.5L = 5.25 m   *Section 4.3.1*

<u>Try 203 × 203 UC 60 S355 (Grade 50)</u>

The Steel Construction Institute 	Job No: **BCB478**		Page **12** of **25**	Rev **A**
	Job Title	**Wind-moment Composite Frames**		
Silwood Park, Ascot, Berks SL5 7QN Telephone: (01344) 623345 Fax: (01344) 622944	Subject	**Internal Columns**		
	Client **SCI**	Made by **JSH**	Date	**Nov. 99**
CALCULATION SHEET		Checked by **AW**	Date	**Dec. 99**

Section is Class 1 plastic   Ref. 18

At L_{Ey} = 3.5 m, P_{cy} = 1720 kN > 1370 kN   OK   Ref. 18

At L_{ex} = 5.25m, P_{cx} = 2105 kN > 1370 kN   OK   Ref. 18

At L = 3.5m, M_{bs} = 226 kNm > 12.6 kNm   OK   Ref. 18

Overall buckling stability check:

$$\frac{F_c}{P_c} + \frac{M_x}{M_{bs}} = \frac{1370}{1720} + \frac{12.6}{226} = 0.85 < 1.0 \quad \text{OK}$$

Section 4.3.5

<u>Dead load plus imposed load plus wind loading</u>

Design load at ULS: F_c = 1.2 × (479 + 437) = 1099 kN   Table D.7

Design moment at ULS: M_x = 1.2 × 10.9 = 13 kNm   Table D.6

Overall buckling check: $\dfrac{1099}{1720} + \dfrac{13}{226} = 0.70 < 1.0$ OK   Section 4.3.5

<u>Dead load plus wind loading</u>

Design load at ULS: F_c = 1.4 × 479 = 671 kN   Table D.7

Design moment at ULS: M_x = 1.4 × 10.9 = 15 kNm   Table D.6

Overall buckling check: $\dfrac{671}{1720} + \dfrac{15}{226} = 0.46 < 1.0$ ∴ OK   Section 4.3.5

<u>Pattern Loading: Dead load plus imposed load plus notional horizontal forces</u>

Design load at ULS: F_c = 1.4 × 479 + 1.6 (0.5 × 437) = 1021 kN   Table D.7

Design moment for ULS: M_x = 12.6 + 1.6 × 40.5 = 77 kNm   Table D.6

Overall buckling check: $\dfrac{1021}{1720} + \dfrac{77}{226} = 0.93 < 1.0$ OK   Section 4.3.5

<u>Use 203 × 203 UC 60 S355 (Grade 50) for internal 3rd and 4th storey columns</u>

The Steel Construction Institute  Silwood Park, Ascot, Berks SL5 7QN Telephone: (01344) 623345 Fax: (01344) 622944 **CALCULATION SHEET**		Job No:	**BCB478**	Page *13* of *25*		Rev *A*
		Job Title	**Wind-moment Composite Frames**			
		Subject	**Internal Columns**			
		Client	*SCI*	Made by	*JSH*	Date *Nov. 99*
				Checked by	*AW*	Date *Dec. 99*

D.7.2 Storey 1

<u>Dead load plus imposed load plus notional horizontal forces</u>

Design load at ULS: $F_c = 1.4 \times 1030 + 1.6 \times 907 = 2893$ kN   *Table D.7*

Design moment at ULS: $M_x = 39.3$ kNm   *Table D.6*

Storey height = 4.5 m

$L_{Ey} = 1.0L = 4.5$ m, $L_{Ex} = 1.5L = 6.75$ m   *Section 4.3.1*

<u>Try 254 × 254 UC 132 S355 (Grade 50)</u>

Section is Class 1 Plastic   *Ref. 18*

At $L_{Ey} = 4.5$m, $P_{cy} = 3715$ kN > 2893 kN   OK   *Ref. 18*

At $L_{Ex} = 6.75$m, $P_{cx} = 4540$ kN > 2893 kN   OK   *Ref. 18*

At $L = 4.5$m, $M_{bs} = 627$ kNm > 39.3 kNm   OK   *Ref. 18*

Overall buckling check: $\dfrac{F_c}{P_c} + \dfrac{M_x}{M_{bs}} = \dfrac{2893}{3715} + \dfrac{39.3}{627} = 0.84 < 1.0$ OK   *Section 4.3.5*

<u>Dead load plus imposed load plus wind loading</u>

Design load at ULS: $F_c = 1.2 (1030 + 907) = 2324$ kN   *Table D.7*

Design moment for ULS: $M_x = 1.2 \times 32.1 = 38.5$ kNm   *Table D.6*

Overall buckling check: $\dfrac{2324}{3715} + \dfrac{38.5}{627} = 0.69 < 1.0$ OK   *Section 4.3.5*

<u>Dead load plus wind loading</u>

Design load at ULS: $F_c = 1.4 \times 1030 = 1442$ kN   *Table D.7*

Design moment at ULS: $M_x = 1.4 \times 32.1 = 44.9$ kNm   *Table D.6*

Overall buckling check: $\dfrac{1442}{3715} + \dfrac{44.9}{627} = 0.46 < 1.0$ OK   *Section 4.3.5*

By inspection, pattern loading is not critical

<u>Use 254 × 254 UC 132 (S355) for internal 1st and 2nd storey columns</u>

D.8 EXTERNAL COLUMN DESIGN

Table D.8   Loading on external columns

Storey	Loading (kN) Dead / Imposed	S_w of Col. (kN)	Total load (kN) Dead	Total load (kN) Imposed	Reduction in imposed load (kN)	Reduced imposed load (kN)
4	101.3 / 40.5	3	104	41	0% / 0	41
3	135.0 / 202.5	3	242	243	10% / 24	219
2	135.0 / 202.5	5	382	446	20% / 89	357
1	135.0 / 202.5	6	523	648	30% / 194	454

N.B.   Values are UNFACTORED

D.8.1 Storey 3

<u>Dead load plus imposed load plus notional horizontal forces</u>

Design load for ULS: F_c = 1.4 × 242 + 1.6 × 219 = 689 kN   Table D.8

Design moment for ULS: M_x

Assume section is 200 mm deep

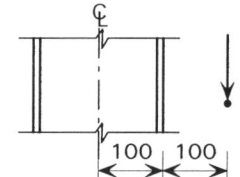

Eccentricity moment   (1.4 × 90 + 1.6 × 135) (0.1 + 0.1) = 68.4 kNm   Table D.6

10% restraint moment   (1.4 × 27) + (1.6 × 40.5) = 102.6 kNm   Table D.6

Sum of moments = 171 kNm

Divide moments equally between upper & lower column lengths = 85.5 kNm   Section 4.3.3

Notional horizontal load induced moment = 6.3 kNm   Table D.6

The Steel Construction Institute  Silwood Park, Ascot, Berks SL5 7QN Telephone: (01344) 623345 Fax: (01344) 622944 **CALCULATION SHEET**	Job No: **BCB478**	Page **15** of **25** Rev **A**
	Job Title **Wind-moment Composite Frame**	
	Subject **External Column Design**	
	Client **SCI**	Made by **JSH** Date **Nov. 99**
		Checked by **AW** Date **Dec. 99**

Total design moment, M_x = <u>91.8 kNm</u>

Column length = 3.5 m

L_{Ey} = 1.0L = 3.5 m, L_{Ex} = 1.5L = 5.25 m   Section 4.3.1

<u>Try 203 × 203 UC 60 S355 (Grade 50)</u>

Section is Class 1 Plastic

At Ley = 3.5 m, P_{cy} = 1720 kN > 689 kN   OK   Ref. 18

At Lex = 5.25 m, P_{cx} = 2105 kN > 689 kN   OK   Ref. 18

At L = 3.5 m, M_{bs} = 226 kNm > 91.8 kNm   OK   Ref. 18

Overall buckling check:

$$\frac{F_c}{P_c} + \frac{M_x}{M_{bs}} = \frac{689}{1720} + \frac{91.8}{226} = 0.81 < 1.0 \therefore OK$$ Section 4.3.5

<u>Dead load plus imposed load plus wind loading</u>

Design load at ULS: F_c = 1.2 (242 + 219 + 2.2) = 556 kN   Table D.8
                                                                Table D.2

Design moment for ULS: M_x

Eccentricity moment   1.2 (90 + 135) (0.1 + 0.1) = 54 kNm   Table D.6

10% restraint moment   1.2 (27 + 40.5) = 81 kNm   Table D.6

Sum of moments = 135 kNm

Divide equally between upper and lower columns = 67.5 kNm

Wind induced moment   1.2 × 5.5 = 6.6 kNm   Table D.6

Total design moment, M_x = <u>74.1 kNm</u>

Overall stability check = $\frac{556}{1720} + \frac{74.1}{226}$ = 0.65 < 1.0   OK   Section 4.3.5

 **The Steel Construction Institute** Silwood Park, Ascot, Berks SL5 7QN Telephone: (01344) 623345 Fax: (01344) 622944		Job No: **BCB478**		Page **16** of **25**		Rev **A**
		Job Title	**Wind-moment Composite Frame**			
		Subject	**External Column Design**			
CALCULATION SHEET		Client **SCI**	Made by **JSH**		Date **Nov. 99**	
			Checked by **AW**		Date **Dec. 99**	

Dead load plus wind loading

By inspection, not critical (stability factor = 0.37)

Use 203 × 203 UC 60 S355 for external 3rd and 4th storey columns

D.8.2 Storey 1

Dead load plus imposed load plus notional horizontal forces

Design load for ULS: F_c = 1.4 × 523 + 1.6 × 454 = 1459 kN   Table D.8

Design moment for ULS: M_x

Assume section is 200 mm deep

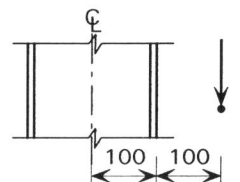

Eccentricity moment   (1.4 × 90 + 1.6 × 135) (0.1 + 0.1) = 68.4 kNm   Table D.6

10% restraint moment   (1.4 × 27) + (1.6 × 40.5)   = 102.6 kNm   Table D.6

Sum of moments   = 171 kNm

Conservatively consider moment to be divided equally   Section 4.3.3
between upper and lower column lengths   = 85.5 kNm

Notional horizontal load induced moment   = 19.6 kNm   Table D.6

Total design moment, M_x = <u>105 kNm</u>

Column length   = 4.5 m

L_{Ey} = 1.0L = 4.5 m, L_{ex} = 1.5 L   = 6.75 m

Try 254 × 254 UC 89 S355 (Grade 50)

Section is Class 1 Plastic

At L_{Ey} = 4.5 m, P_{cy} = 2455 kN > 1459 kN   OK

At L_{Ex} = 6.75 m, P_{cx} = 3005 kN > 1459 kN   OK

At   L   = 4.5 m, M_{bs} = 408 kNm > 105 kNm   OK

The Steel Construction Institute  Silwood Park, Ascot, Berks SL5 7QN Telephone: (01344) 623345 Fax: (01344) 622944 **CALCULATION SHEET**	Job No: **BCB478**		Page **17** of **25**	Rev **A**
	Job Title	**Wind-moment Composite Frame**		
	Subject	**External Column Design**		
	Client **SCI**	Made by **JSH**	Date	**Nov. 99**
		Checked by **AW**	Date	**Dec. 99**

Overall buckling check:

$$\frac{F_c}{P_c} + \frac{M_x}{M_{bs}} = \frac{1459}{2455} + \frac{105}{408} = 0.85 < 1.0 \quad \text{OK}$$

Section 4.3.5

Dead load plus imposed load plus wind loading

Design load for ULS: $F_c = 1.2 (523 + 454 + 10.9)$ $= 1185\ kN$

Design moment for ULS: M_x

Eccentricity moment	$1.2 (90 + 135) (0.1 + 0.1)$	$= 54\ kNm$
10% restraint moment	$1.2 (2.7 + 40.5)$	$= 81\ kNm$
Sum of moments		$= 135\ kNm$
Divide equally between upper and lower columns		$= 67.5\ kNm$
Wind induced moment	1.2×16	$= 19\ kNm$
Total design moment		$= \underline{86.5\ kNm}$

Table D.6 (for eccentricity, 10% restraint, wind induced moments)

Overall stability check $= \dfrac{1185}{2455} + \dfrac{86.5}{408} = 0.69 < 1.0$ OK

Section 4.3.5

Dead load plus wind loading

By inspection, not critical (stability factor $= 0.44$)

Use 254×254 UC 89 S355 for external 1st and 2nd storey columns

The Steel Construction Institute  Silwood Park, Ascot, Berks SL5 7QN Telephone: (01344) 623345 Fax: (01344) 622944 **CALCULATION SHEET**	Job No: **BCB478**		Page **18** of **25**	Rev **A**
	Job Title	**Wind-moment Composite Frame**		
	Subject	**Connection Loads**		
	Client **SCI**	Made by **JSH**	Date **Nov. 99**	
		Checked by **AW**	Date **Dec. 99**	

D.9 CONNECTION

D.9.1 Connection loading

Calculations are given for connections at 1st floor level only, as an example of how the design should be carried out.

<u>Dead load plus imposed load plus notional horizontal forces</u>

Maximum hogging moment: $= (1.4 \times 27) + (1.6 \times 40.5) + 30.4$ Table D.5
(moment from 10% beam restraint   Table D.6
and notional horizontal forces) $= 133\ kNm$

Minimum hogging moment: $= (1.4 \times 27) + (1.6 \times 40.5) - 30.4$ Table D.5
(moment from 10% beam restraint   Table D.6
and notional horizontal forces) $= 72\ kNm$

Design shear at ULS: $F_v = (1.4 \times 90) + (1.6 \times 135) + \dfrac{30.4}{4.5}$ Table D.6
   Table D.5
(Shear from beam reactions
and notional horizontal forces) $= 349\ kN$

<u>Dead load plus imposed load plus wind loading</u>

Maximum hogging moment: $= 1.2\,(27 + 40.5 + 24.8)$ Table D.6
(moment from 10% beam restraint   Table D.3
and wind loading) $= 111\ kNm$

Minimum hogging moment: $= 1.2\,(27 + 40.5 - 24.8)$ Table D.6
(moment from 10% beam restraint   Table D.3
and wind loading) $= 51\ kNm$

Design shear for ULS: $F_v = 1.2\,(90 + 135 + \dfrac{24.8}{4.5})$ Table D.6
   Table D.3
(shear from beam reactions
and wind loading) $= 277\ kN$

<u>Dead load plus wind loading</u>

Maximum hogging moment: $= 1.4\,(27 + 24.8)$ Table D.6
(moment from 10% beam restraint   Table D.3
and wind loading) $= 73\ kNm$

Minimum hogging moment: $= 1.4\,(27 - 24.8)$ Table D.6
(moment from 10% beam restraint   Table D.3
and wind loading) $= 3.1\ kNm$

The Steel Construction Institute  Silwood Park, Ascot, Berks SL5 7QN Telephone: (01344) 623345 Fax: (01344) 622944		Job No:	**BCB478**	Page **19** of **25**		Rev **A**
		Job Title	**Wind-moment Composite Frame**			
		Subject	**Connection Loads**			
		Client **SCI**		Made by **JSH**	Date	**Nov. 99**
CALCULATION SHEET				Checked by **AW**	Date	**Dec. 99**

The connection does not go into sagging when wind direction reverses. Therefore, the connection should be designed for the largest hogging moment of 133 kNm, nominal sagging resistance will suffice (see Section 4.4.1).

Design 1st floor connections for $\quad M = 133\ kNm$

and $\quad\quad\quad\quad\quad\quad\quad\quad F_v = 349\ kN$

Check rebar area for greatest difference between maximum and minimum hogging moment $\quad = 73 - 3.1 \quad = 69\ kNm$

D.9.2 Connection design

<u>Internal Connection</u>

Beam size 457×191 UB 98 — Sheet 10

Try connection with rebar Option A (4 No 16ϕ bars) and 1 row of M20 8.8 bolts, 200×12 S275 end plate from page C2. — Appendix C

Connection is not acceptable with S355 beam.

Therefore, try rebar Option B (6 No 16ϕ bars)

Moment capacity (beam side) $\quad = 362\ kNm > 133\ kNm \quad$ OK — Appendix C

Vertical shear capacity (with shear row) $= 442\ kN > 349\ kN \quad$ OK — Appendix C

Note that from page C2 bars will need to be X16, specified to be able to achieve a strain at maximum load of at least 10%.

Check that rebar area satisfies upper limit:

Area $\quad\quad\quad = 1210\ mm^2$

Limit $\quad A_L \quad = 1.5 b_c d_s f_{cu} / \mu f_y$ — Section B.2.2

With $\quad \mu \quad\quad = 1 - M_{low}/M_{high} \quad = 1 - (362 - 69)/362 \quad = 0.19$

So $\quad A_L \quad = 1.5 \times 256 \times 90 \times 30 / 0.19 \times 460 \quad = 11863\ mm^2 \quad$ OK

Column $\quad\quad = 254 \times 254$ UC 132 (S355) — Sheet 13

From tables: Column does not require compression stiffening $\quad$ OK

Panel shear capacity $= 892\ kN$ — Appendix C

The Steel Construction Institute 	Job No: **BCB478**		Page **20** of **25**	Rev **A**
	Job Title	**Wind-moment Composite Frame**		
Silwood Park, Ascot, Berks SL5 7QN Telephone: (01344) 623345 Fax: (01344) 622944	Subject	**Connection Loads**		
	Client **SCI**	Made by **JSH**	Date	**Nov. 99**
CALCULATION SHEET		Checked by **AW**	Date	**Dec. 99**

As there are no sagging moments at the connections, conservatively assume that

Panel shear $= F_{reinf} + F_{bolts} = 529 + 208 = 737$ kN < 892 kN   OK   Appendix C

Use rebar Option B and 1 row of M20 8.8 bolts and 200 × 12 S275 end plate

External connection

Column $= 254 \times 254$ UC 89 (S355)   Sheet 17

There is no rebar anchorage around column therefore the connection must be designed as a bare steel wind-moment connection

Try connection with 2 rows of M20 8.8 bolts, 200 × 12 (S275) extended end plate

Moment capacity (beam side) $= 141$ kNm > 133 kNm   OK   Ref. 1

Vertical shear capacity (no shear row) $= 515$ kN > 349 kN   OK   Ref. 1

Column does not require compression stiffening

Panel shear capacity $= 566$ kN

Panel shear $= \sum F_{bolts} = 124 + 208 = 332$ kN < 566 kN   OK

Use 2 rows of M20 8.8 bolts, 200 × 12 (S275) extended end plate

(3rd floor and roof connections should be designed in the same manner.)

The Steel Construction Institute 	Job No: **BCB478**		Page **21** of **25**	Rev **A**
	Job Title	**Wind-moment Composite Frames**		
Silwood Park, Ascot, Berks SL5 7QN Telephone: (01344) 623345 Fax: (01344) 622944	Subject	**Serviceability Limit State**		
	Client **SCI**	Made by **JSH**	Date	**Nov. 99**
CALCULATION SHEET		Checked by **AW**	Date	**Dec. 99**

D.10 SERVICEABILITY LIMIT STATE - SWAY DUE TO WIND

Sway deflections can be calculated using any recognised method. The method used in the design example is a simplified procedure developed by Wood and Roberts [20].

The actual frame is replaced by a substitute beam-column frame. The basis of the substitute frame is that:

(i) For horizontal loading on the actual frame, the rotations of all joints at any one level are approximately equal, and

(ii) Each beam restrains a column at both ends.

The total stiffness K_b of a beam in the substitute frame is obtained from a summation over all the beams in the actual frame at the level being considered.

The total stiffness K_c of a column in the substitute frame is obtained by a summation over all the columns in the actual frame at the level being considered.

In the simplified method of Wood and Roberts, the sway of a storey is dependent partly on stiffness distribution coefficients calculated for the substitute frame.

To allow for continuity of columns in a multi-storey structure, it is recognised that each floor beam restrains column lengths above and below its own level. This is reflected in the form of the distribution coefficients.

The stiffness distribution coefficients enable a non-dimensional sway index $\bar{\varphi}$ to be determined from the chart given below (Figure B.11). By definition:

$$\bar{\varphi} = \frac{\Delta/h}{Fh/(12EK_c)}$$

where
- Δ/h *is the sway angle of the storey being considered*
- F *is the total wind shear on the column of the substitute frame*
- E *is Young's modulus of elasticity (205 kN/mm^2)*

Values of k_t and k_b for use with Figure D.8 are defined and calculated in Table D.11.

The Steel Construction Institute 	Job No:	BCB478		Page	22	of	25	Rev	A
Silwood Park, Ascot, Berks SL5 7QN Telephone: (01344) 623345 Fax: (01344) 622944	Job Title	\multicolumn{7}{l	}{*Wind-moment Composite Frames*}	
	Subject	\multicolumn{7}{l	}{*Serviceability Limit State*}	
CALCULATION SHEET	Client	*SCI*		Made by		*JSH*	Date		*Nov. 99*
				Checked by		*AW*	Date		*Dec. 99*

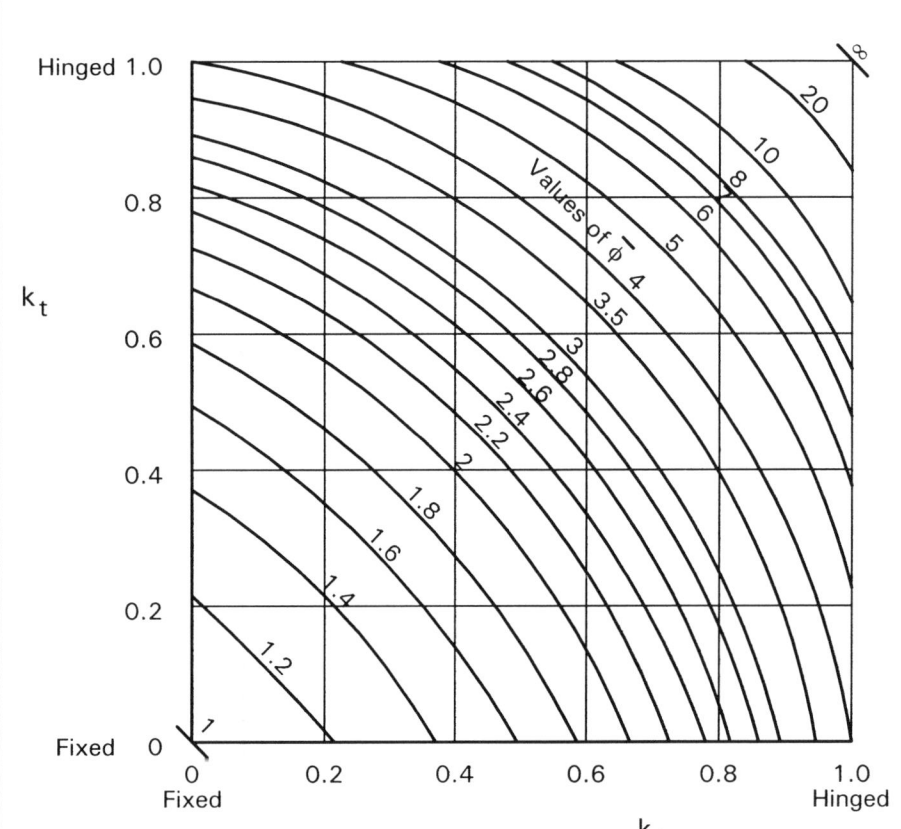

Figure D.8   Sway index ϕ

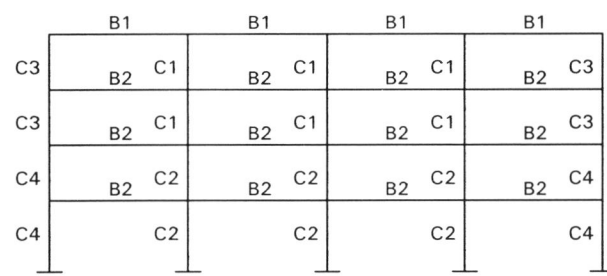
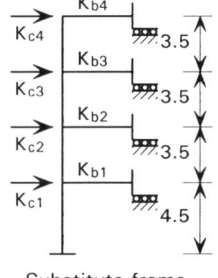

Substitute frame

Figure D.9   Frame model for serviceability limit state calculations

Members:

$B1$ = 356×171 UB $57 + 140$ slab
$B2$ = 457×191 UB $98 + 140$ slab

$C1$ = 203×203 UC 60

			Job No:	**BCB478**	Page	**23** of **25**	Rev	**A**
The Steel Construction Institute			Job Title	*Wind-moment Composite Frames*				
			Subject	*Serviceability Limit State*				
Silwood Park, Ascot, Berks SL5 7QN Telephone: (01344) 623345 Fax: (01344) 622944			Client	*SCI*	Made by	*JSH*	Date	*Nov. 99*
CALCULATION SHEET					Checked by	*AW*	Date	*Dec. 99*

$C2 \ = \ 254 \times 254 \ UC \ 132$
$C3 \ = \ 203 \times 203 \ UC \ 60$
$C4 \ = \ 254 \times 254 \ UC \ 89$

<u>*Equivalent beam stiffness*</u>

Roof Beam: $I_g = 49927 \ cm^4$ $\qquad I_n = 31113 \ cm^4$ Ref. 3
(values calculated using method give in Ref. 3, Appendix B, B.3)

$$I_{eq} = \frac{7.5 I_g I_n^2}{9 I_n^2 + 2 I_g I_n} = \frac{7.5 \times 49927 \times 31113^2}{9 \times 31113^2 + 2(49927 \times 31113)} = 30669 \ cm^4$$

Section 5.2.2

Floor Beam: $I_g = 110424 \ cm^4$ $\qquad I_n = 70697 \ cm^4$ Ref. 3

$$I_{eq} = \frac{7.5 I_g I_n^2}{9 I_n^2 + 2 I_g I_n} = \frac{7.5 \times 110424 \times 70697^2}{9 \times 70697^2 + 2(110424 \times 70697)} = 68310 \ cm^4$$

Section 5.2.2

<u>*Stiffness calculations for substitute frame*</u>

Table D.9   Beam stiffness

Storey	I_b (cm⁴)	L_b (cm)	$K_b = 3 \ \Sigma I_b / L_b$	K_b (cm³)
4	30669	900	3 × 4 × 30669/900	408.9
3	68310	900	3 × 4 × 68310/900	910.8
2	68310	900	3 × 4 × 68310/900	910.8
1	68310	900	3 × 4 × 68310/900	910.8

Table D.10   Column stiffness

Storey	I_b (cm⁴)	L_b (cm)	h (cm)	$K_b = 3 \ \Sigma I_b / L_b$	K_b (cm³)
4	6125	6125	350	5 × 6125/350	87.5
3	6125	6125	350	5 × 6125/350	87.5
2	14270	22530	350	(2 × 14270 + 3 × 22530) / 350	274.7
1	14270	22530	450	(2 × 14270 + 3 × 22530) / 450	213.6

The Steel Construction Institute  Silwood Park, Ascot, Berks SL5 7QN Telephone: (01344) 623345 Fax: (01344) 622944 **CALCULATION SHEET**	Job No: **BCB478**		Page **24** of **25**	Rev **A**
	Job Title	**Wind-moment Composite Frames**		
	Subject	**Serviceability Limit State**		
	Client **SCI**	Made by **JSH**	Date	**Nov. 99**
		Checked by **AW**	Date	**Dec. 99**

Stiffness distribution coefficients

Table D.11   Joint stiffness coefficients

Storey	$K_r = \dfrac{K_c + K_u}{K_c + K_u + K_{bt}}$	K_t	$K_b = \dfrac{K_c + K_L}{K_c + K_L + K_{bb}}$	K_b
4	$\dfrac{87.5 + 0}{87.5 + 0 + 408.6}$	0.18	$\dfrac{87.5 + 87.5}{87.5 + 87.5 + 910.8}$	0.16
3	$\dfrac{87.5 + 87.5}{87.5 + 87.5 + 910.8}$	0.16	$\dfrac{87.5 + 274.7}{87.5 + 274.7 + 910.8}$	0.28
2	$\dfrac{274.7 + 87.5}{274.7 + 87.5 + 910.8}$	0.28	$\dfrac{274.7 + 213.6}{274.7 + 213.6 + 910.8}$	0.35
1	$\dfrac{213.6 + 274.7}{213.6 + 274.7 + 910.8}$	0.35	Nominal fixed base	0

Where: K_u is the stiffness of the column above the storey
         K_l is the stiffness of the column below the storey
         K_{bt} is the stiffness of the beam above the storey
         K_{bb} is the stiffness of the beam below the storey

Sway deflections

Table D.12   Sway deflections for a rigid frame

Storey	K_t	K_b	$\bar{\phi}$	$F(kN)$	$\dfrac{\Delta}{h} = \dfrac{Fh\bar{\phi}}{12EK_c}$	$\dfrac{\Delta}{h}$	Δ (mm)
4	0.18	0.16	1.31	10	$\dfrac{(10 \times 350 \times 1.31)}{(12 \times 20500 \times 87.5)}$	1/4694	0.7
3	0.16	0.28	1.44	25	$\dfrac{(25 \times 350 \times 1.44)}{(12 \times 20500 \times 87.5)}$	1/1708	2
2	0.28	0.35	1.71	40	$\dfrac{(40 \times 350 \times 1.71)}{(12 \times 20500 \times 274.7)}$	35800	1.2
1	0.35	0	1.35	57	$\dfrac{(57 \times 450 \times 1.35)}{(12 \times 20500 \times 213.6)}$	1/1517	3
					TOTAL	1/2142	7

The Steel Construction Institute  Silwood Park, Ascot, Berks SL5 7QN Telephone: (01344) 623345 Fax: (01344) 622944	Job No: **BCB478**	Page **25** of **25** Rev **A**
	Job Title **Wind-moment Composite Frames**	
	Subject **Serviceability Limit State**	
CALCULATION SHEET	Client **SCI**	Made by **JSH** Date **Nov. 99**
		Checked by **AW** Date **Dec. 99**

Allowance for connection flexibility

The deflections calculated treating the frame as rigid-jointed are increased the sway amplification factor to make an appropriate allowance for connection flexibility (see Section 5.2.3)

Sway amplification factor for frame with non-composite external connections = 1.6

Section 5.2.3
Table 3

Table D.13 *Sway deflections allowing for connection flexibility*

Storey	Rigid frame Δ (mm)	Wind-moment frame 1.6Δ (mm)	Wind-moment frame $1.6\Delta/h$ (mm)	LIMIT	Check
4	0.7	1.1	1/3182	1/300	OK
3	2	3.2	1/1094	1/300	OK
2	1.2	1.9	1/1842	1/300	OK
1	3	4.8	1/938	1/300	OK
Total	7	11	1/1364	1/300	OK

Table D.12

Section 5.2.4

Frame design is acceptable for wind deflections under SLS loading.

Typeset and page make-up by The Steel Construction Institute, Ascot, Berks. SL5 7QN
Printed and bound by Alden Press, Osney Mead, Oxford, OX2 0EF
1500 - 6/00 (BCB478)